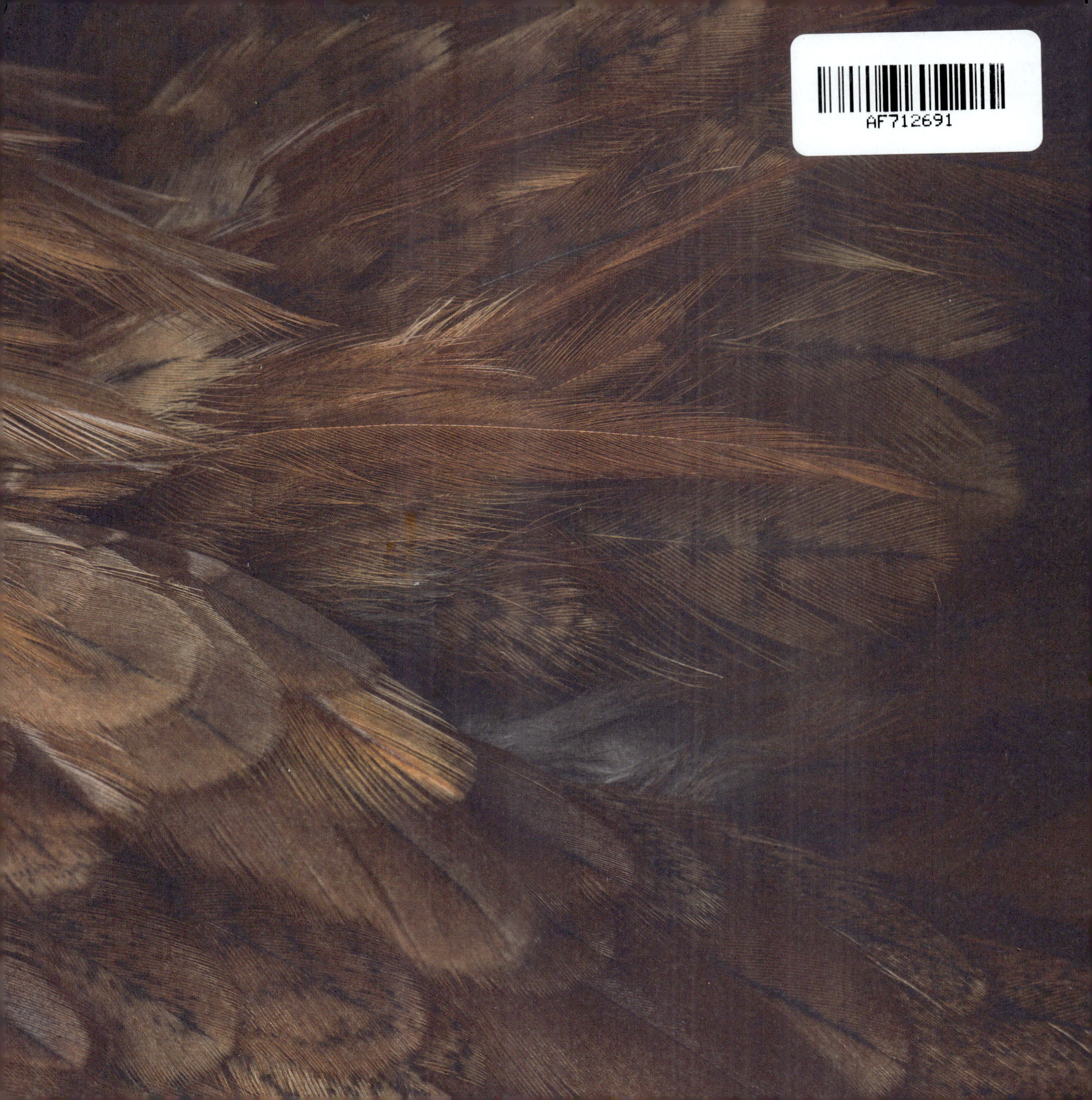

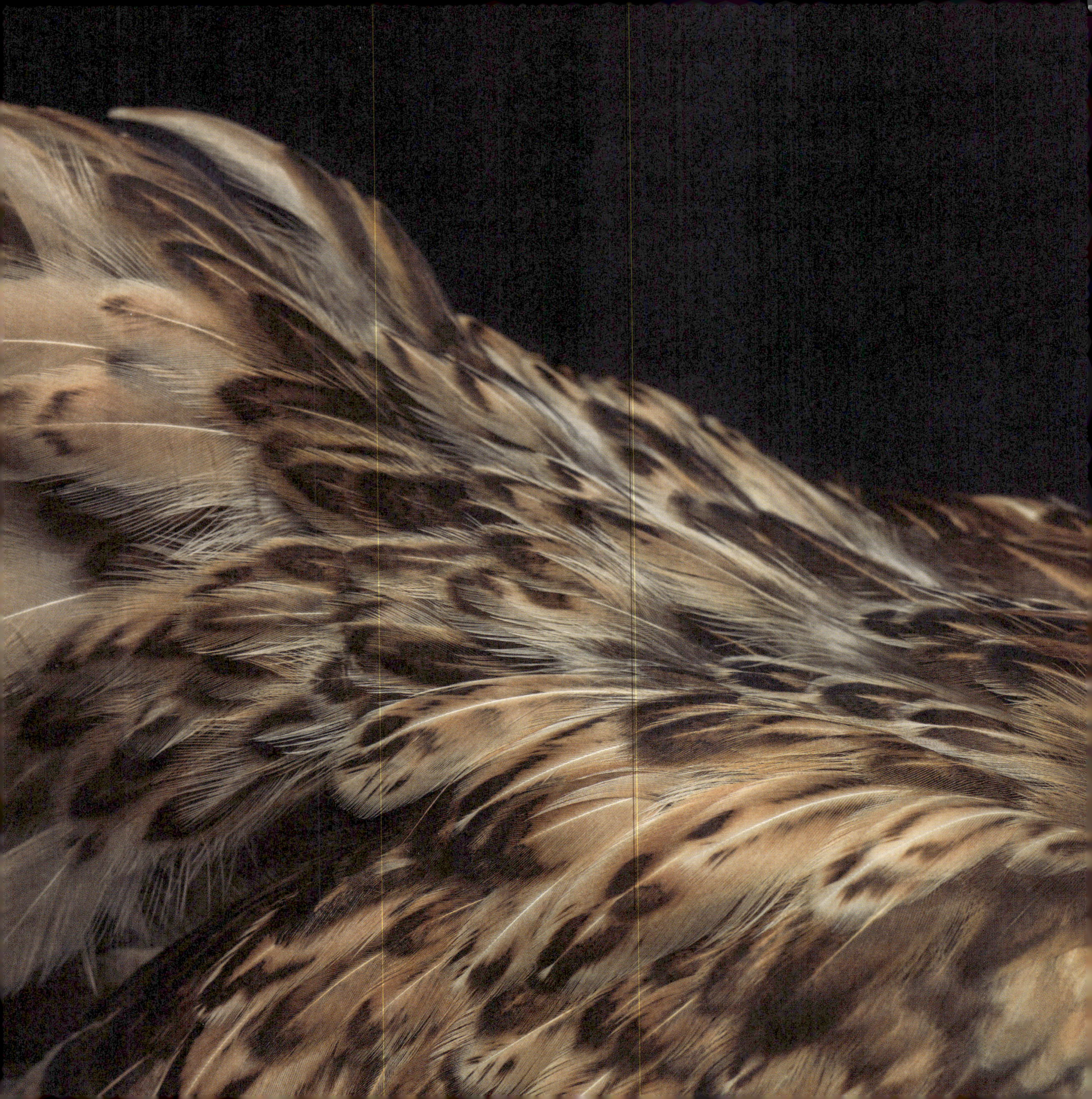

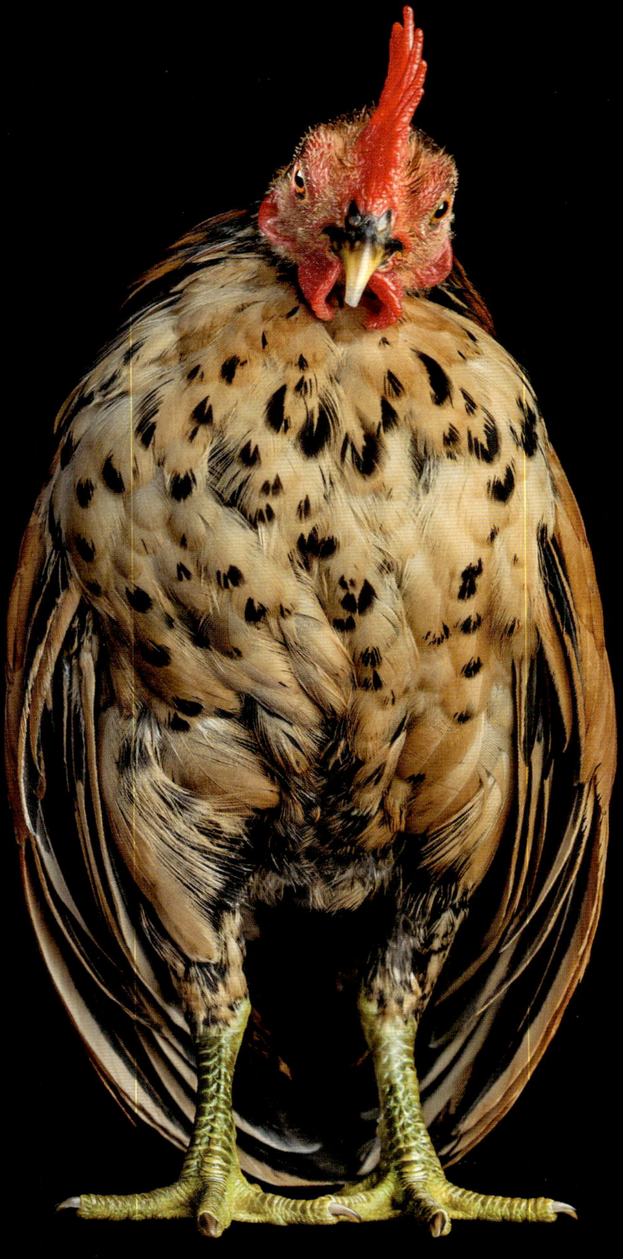

To my baby nephew Dexter:

Love all creatures great and small, even the crazy-looking one

Chickens

ERNEST GOH

Abrams, New York
in association with PQ Blackwell

NEITHER FISH NOR **FOWL**

My interest in animals was nurtured while growing up in my grandmother's *kampung* (village) in 1980s Singapore—a very different time and place from today's concrete jungle. As my boyhood ended I left that life behind, but rediscovered my passion for animals when I worked on my first book, *The Fish Book*. Photographing fish was a real eye-opener. I found each of them to be different, and not just in terms of shape, size, and species. My encounter with each was different, and I came to know them all as individuals. Some were livelier than others, some were more camera-shy, but all had facial expressions of some form—elated, grumpy, excited, or confused. From behind the camera, they seemed to be showing a gamut of emotions.

Since then, I've photographed orangutans in the forests of Sumatra and come face-to-face with snakes in the home of a reptile researcher. These experiences made me realize how our perception of animals can be changed simply by getting to know them as individuals.

For this book on cocks and other fowls, I again chose portraiture to capture their liveliness by setting up a photo studio on location. The main motivation was to discover not what they were but who they were.

Many of the chickens pictured in this book are a breed known as Ayam Serama. *Ayam* is the Malay word for "chicken" and *Serama* is the name given to this particular breed, which is considered to be the smallest in the world. The Serama originated from Malaysia through the crossbreeding of Japanese and Malaysian bantams, but traditional stories tell of a Thai king who gave some small chickens to a Malay sultan. The modern breed was thus named Serama, after *Rama*, the title of kings in Thailand.

The breeding of Serama as pets has become more popular in the last decade. Serama breeding clubs have emerged in Indonesia, Thailand, the USA, and some European countries including the United Kingdom and France, but Malaysia remains the hub of this cultural phenomenon. Serama are often mistaken for fighting cocks—a common misconception because most people tend to associate any rooster competition with the cruel sport of cockfighting. Serama competitions, however, are in fact a beauty contest. Competitions are held almost every week in different villages in Malaysia, where judges sit around a square table to inspect each chicken for a few minutes to determine a champion specimen in its weight class based on its stance, showmanship, and physical assets such as wings, tail, and comb.

When I first started photographing Serama it was strange to see chickens showily display themselves like peafowls, but after observing them for a while I started to understand the uniqueness of this animal, and why Serama owners often regard their chickens as warriors ready for battle. In certain stances, the Serama stands with its head lifted skyward, chest puffed up, and wings pointed downward in a vertical manner, resembling a proud soldier standing at attention. The Serama can also effortlessly prance, strut, and parade, in full pomp and pageantry, and from the perspective of a photographer, the Serama seems more runway model than bird.

Pet owners tend to anthropomorphize their pets and even treat them as members of the family. This is true of many cat and dog owners. In fact, this experience of closeness is prevalent across all human-animal relationships, whether the animals are domesticated or not.

British broadcaster and naturalist Sir David Attenborough once said, "People are not going to care about animal conservation unless they think that animals are worthwhile." I have met farmers who name and pat their pigs, and conservationists who hug and talk to their research subjects. They form long-term relationships with the animals that see both human and animal benefit from an exchange of affection. Perhaps to the human, these benefits are emotional, academic, or economic. Whatever the reason, they all think their animals are worthy of care and protection and that they are not merely animals, or in my case, not just a fish or a fowl.

ERNEST GOH

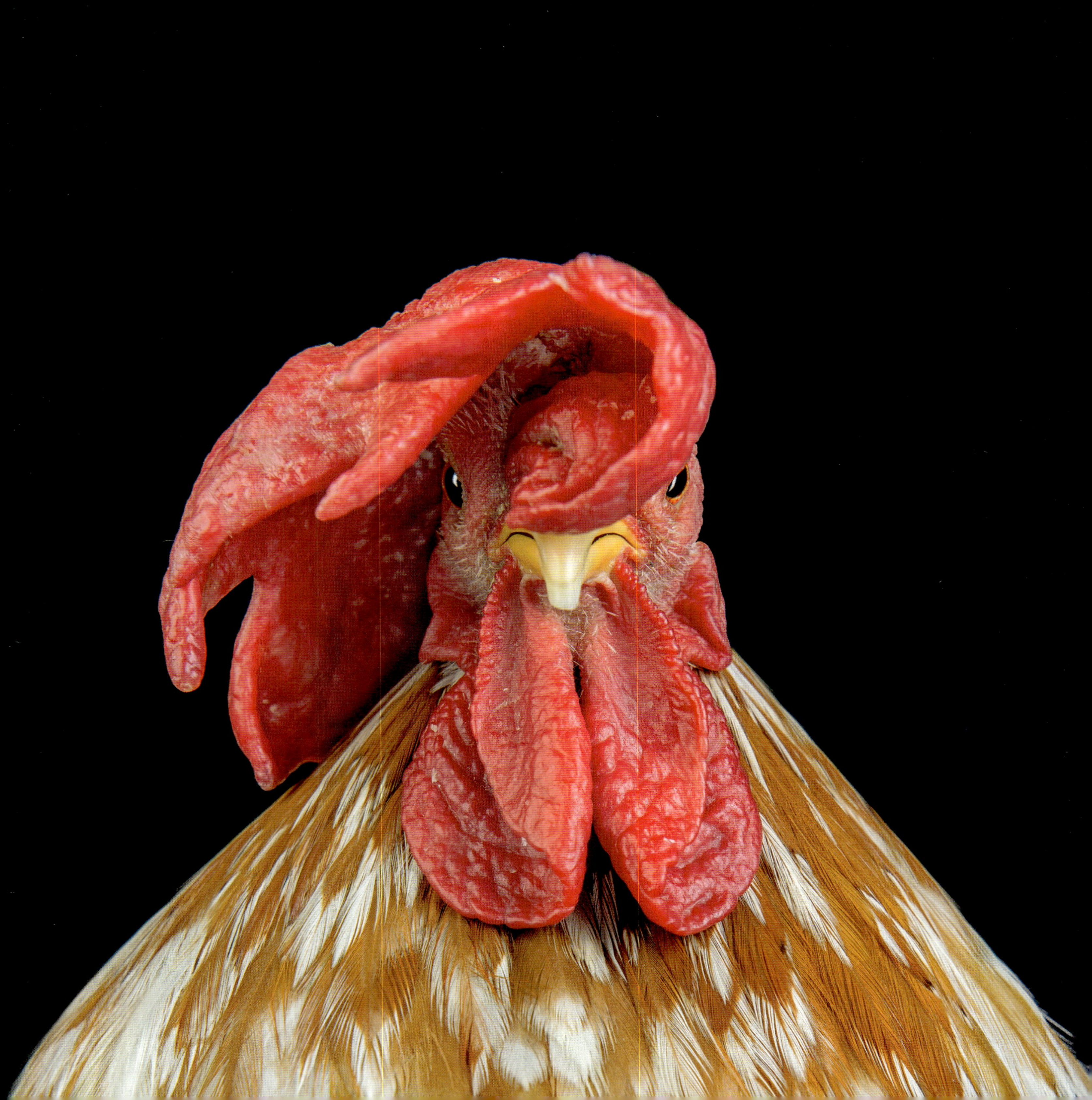

THERE AIN'T NOBODY HERE BUT US

CHICKENS.

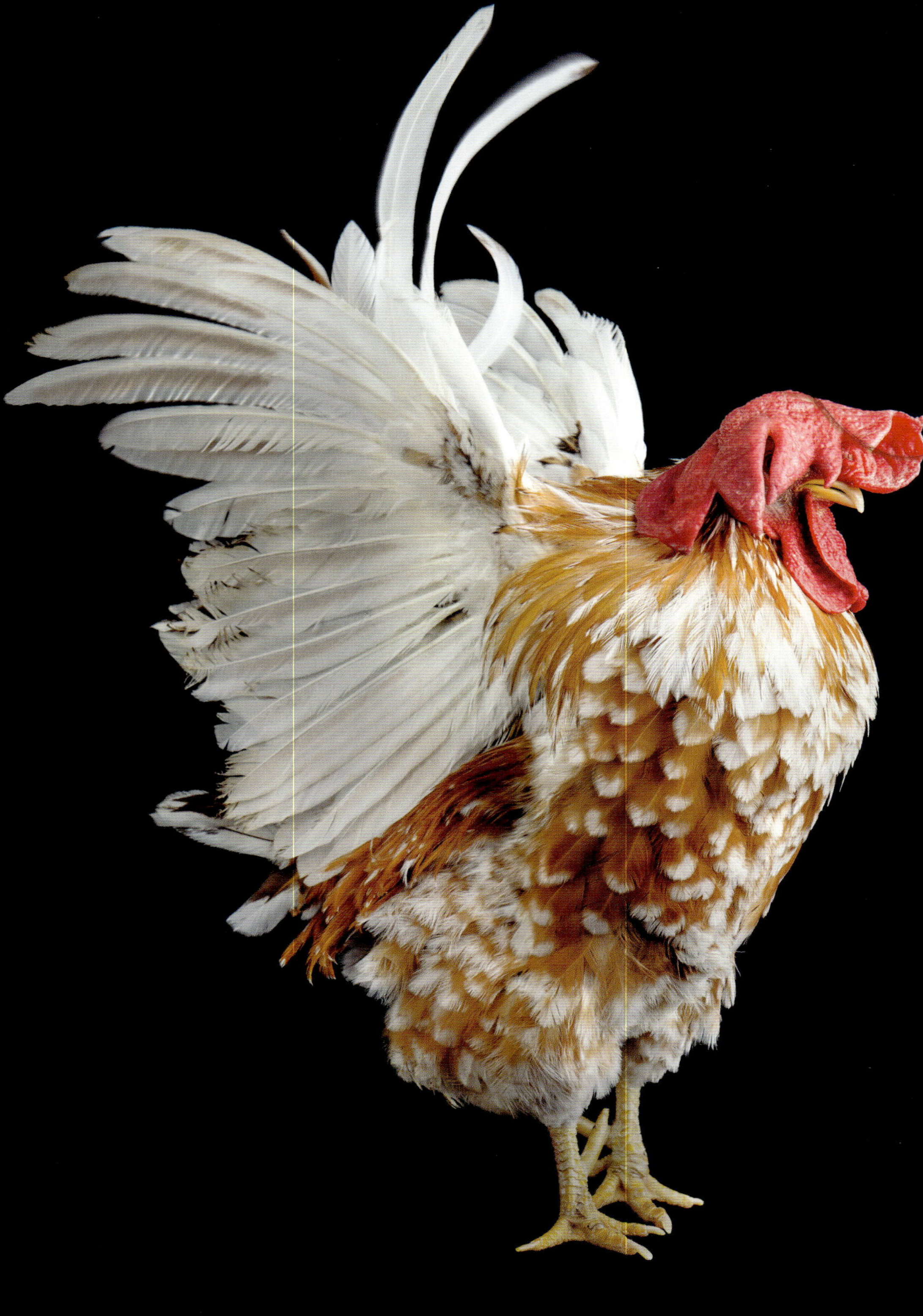

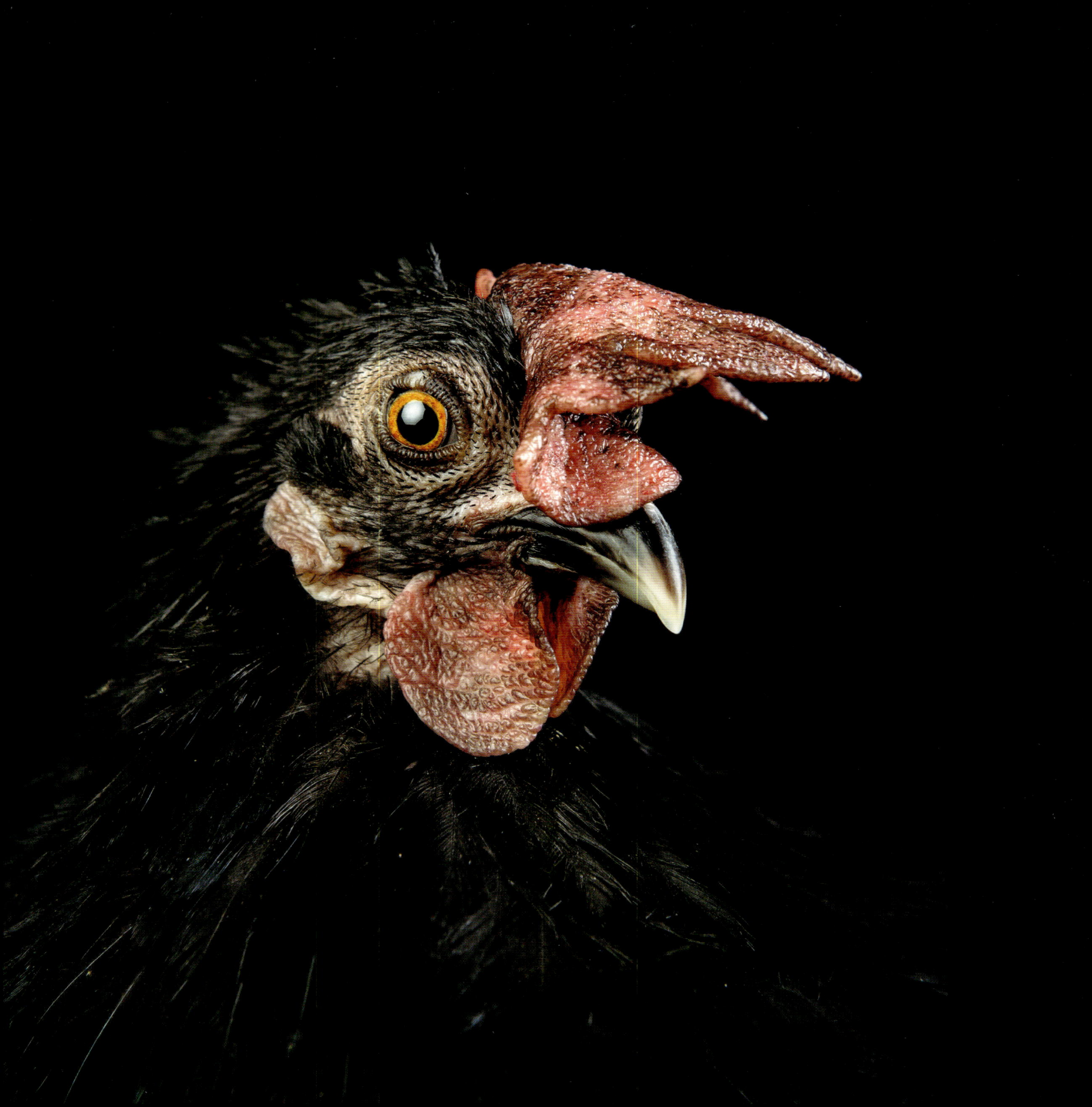

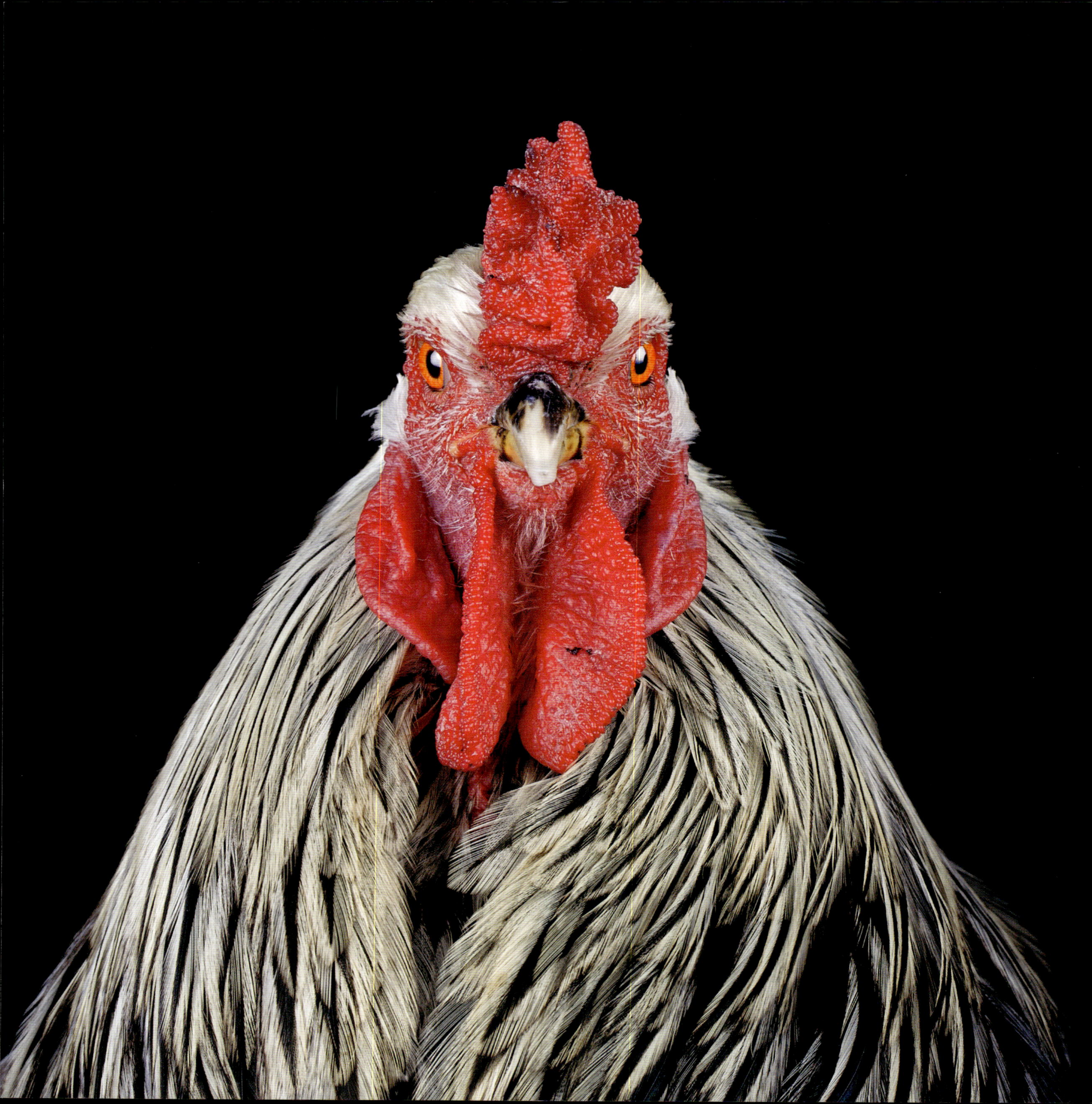

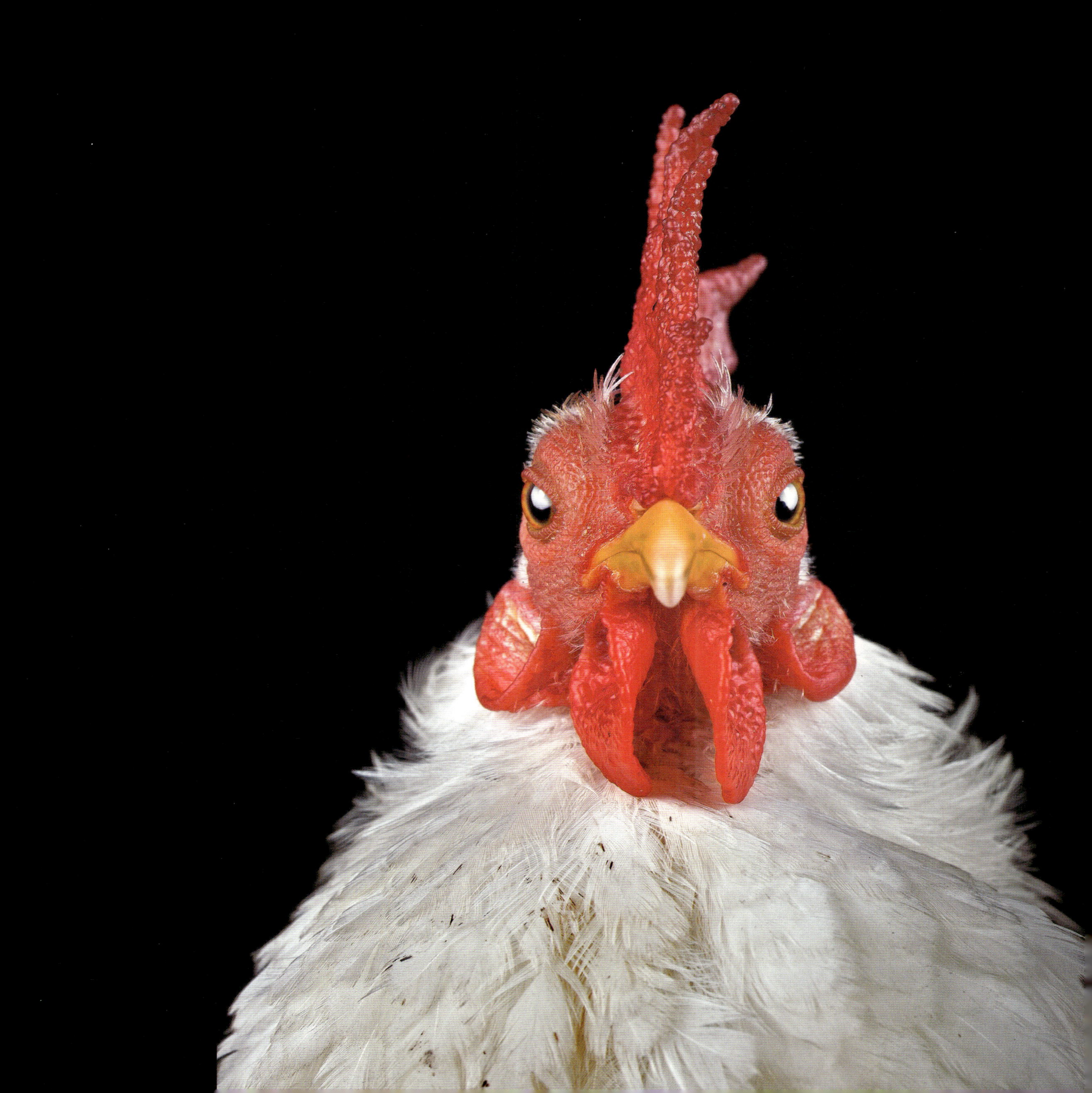

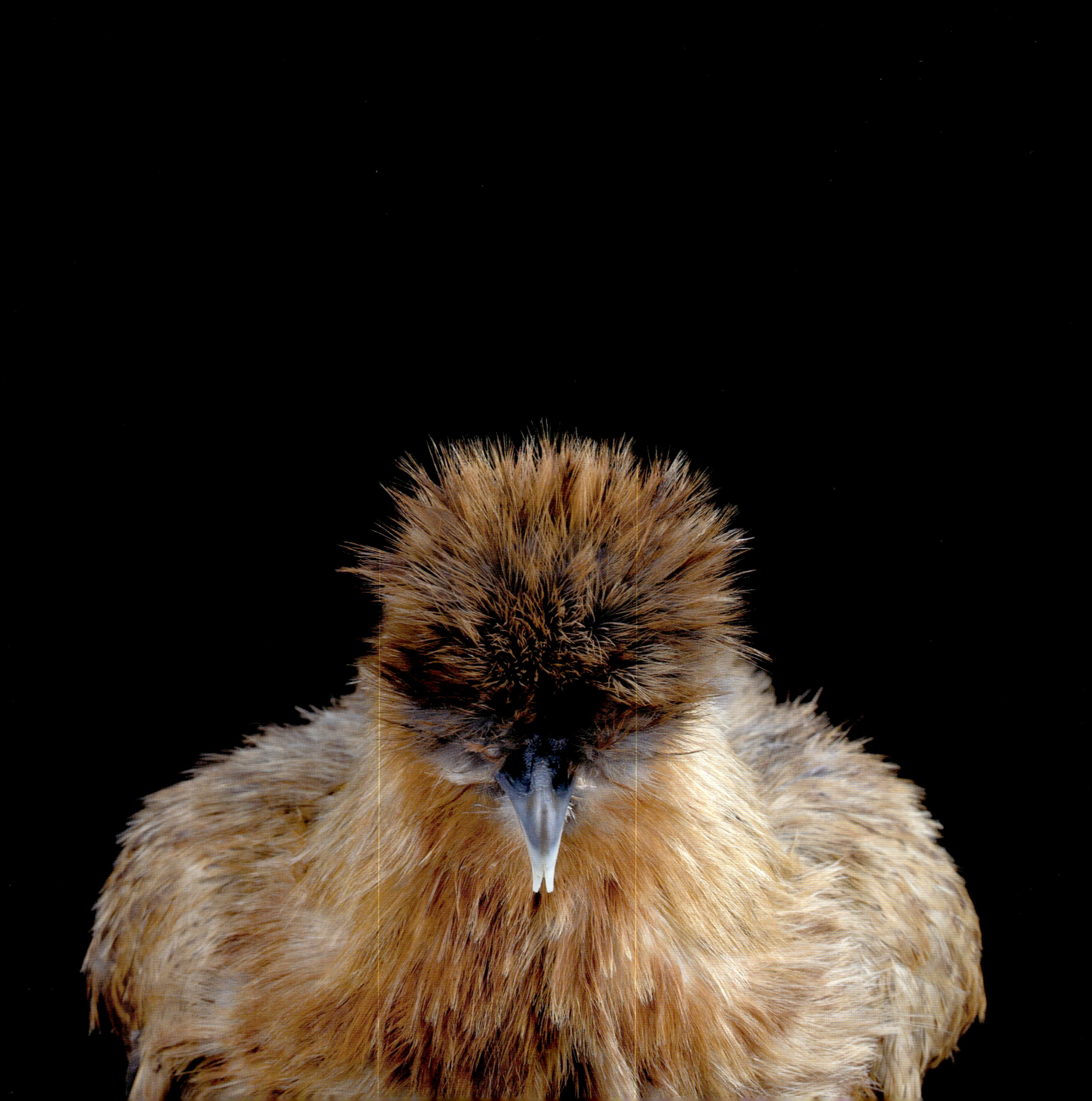

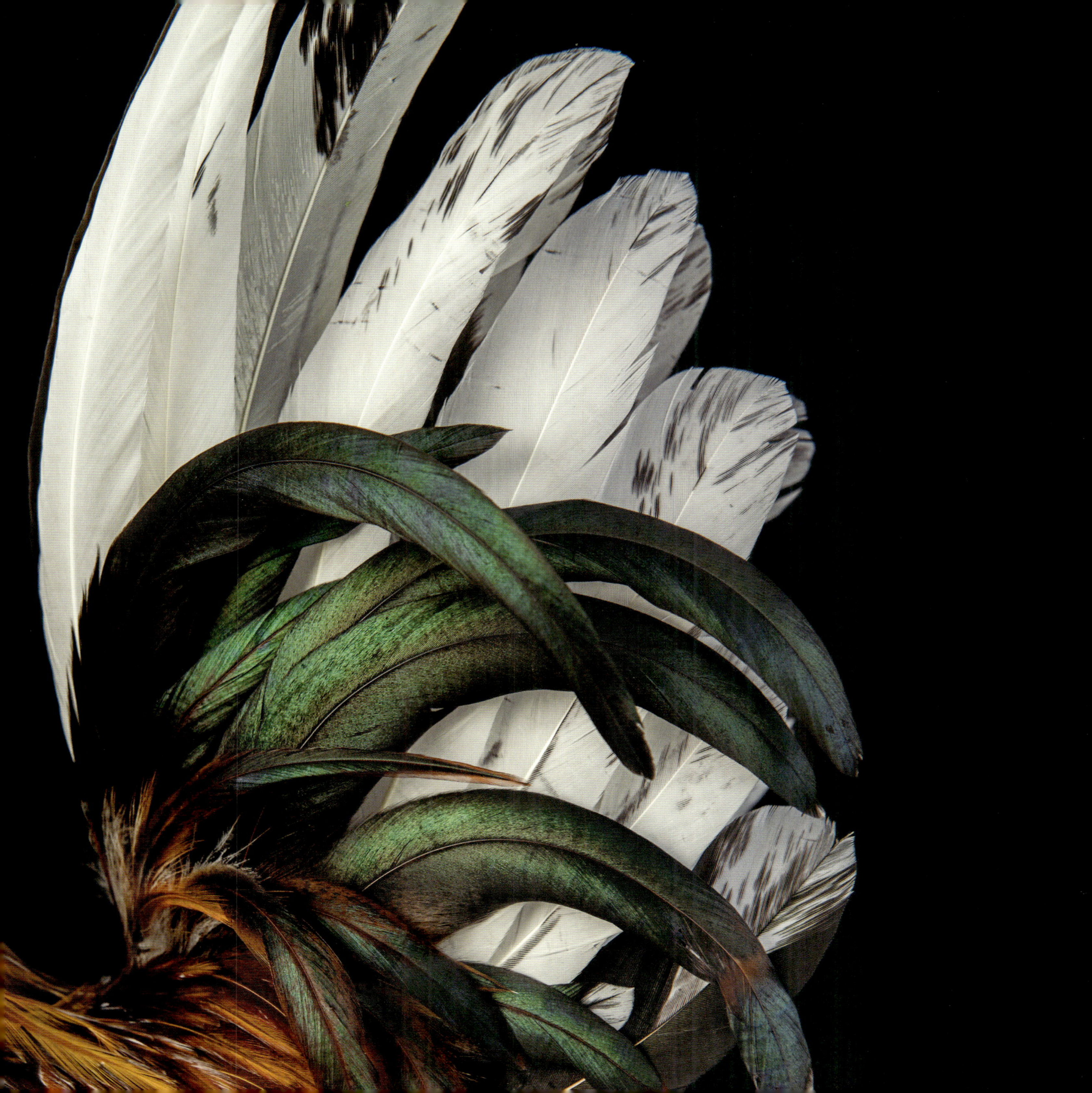

COCK O'

THE WALK...

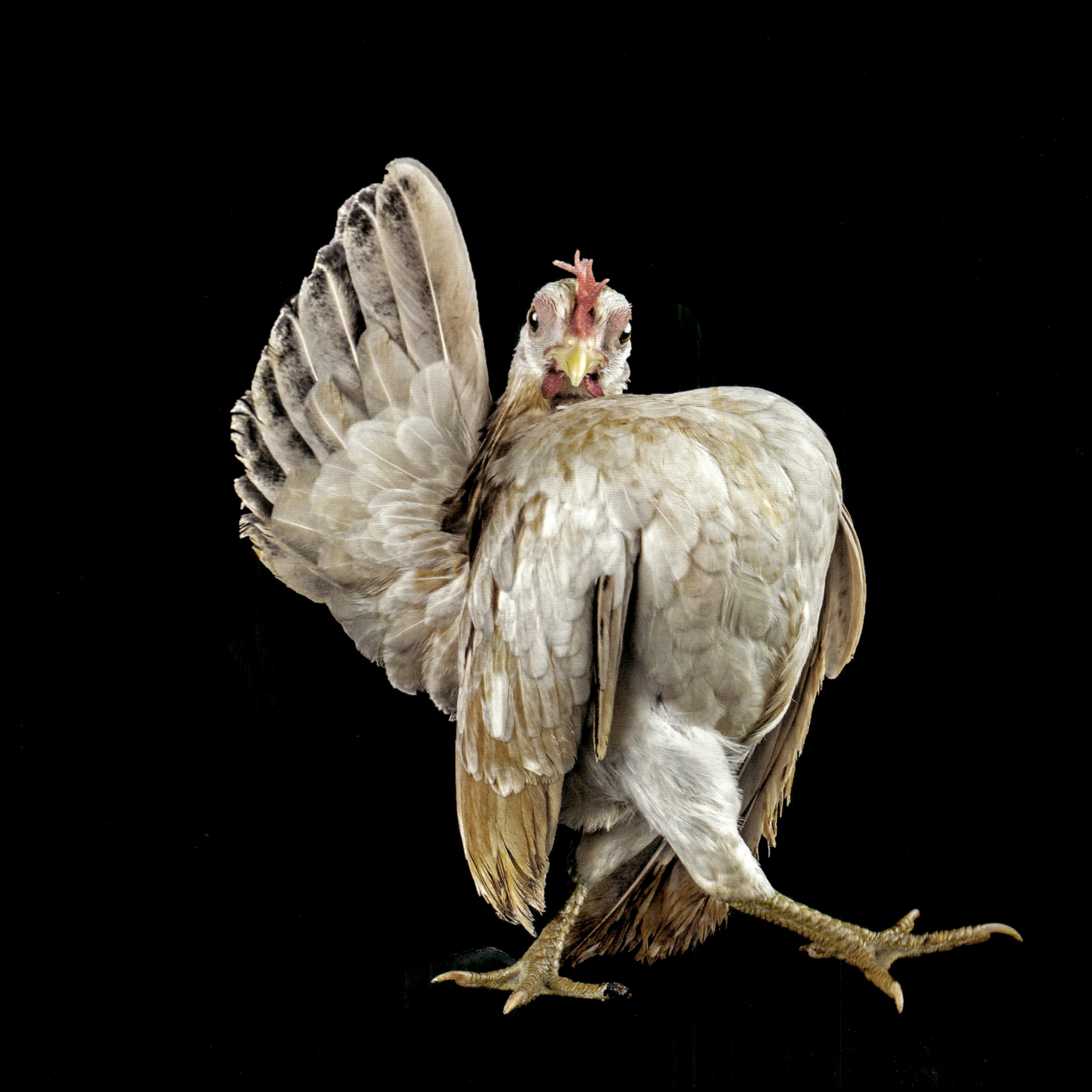

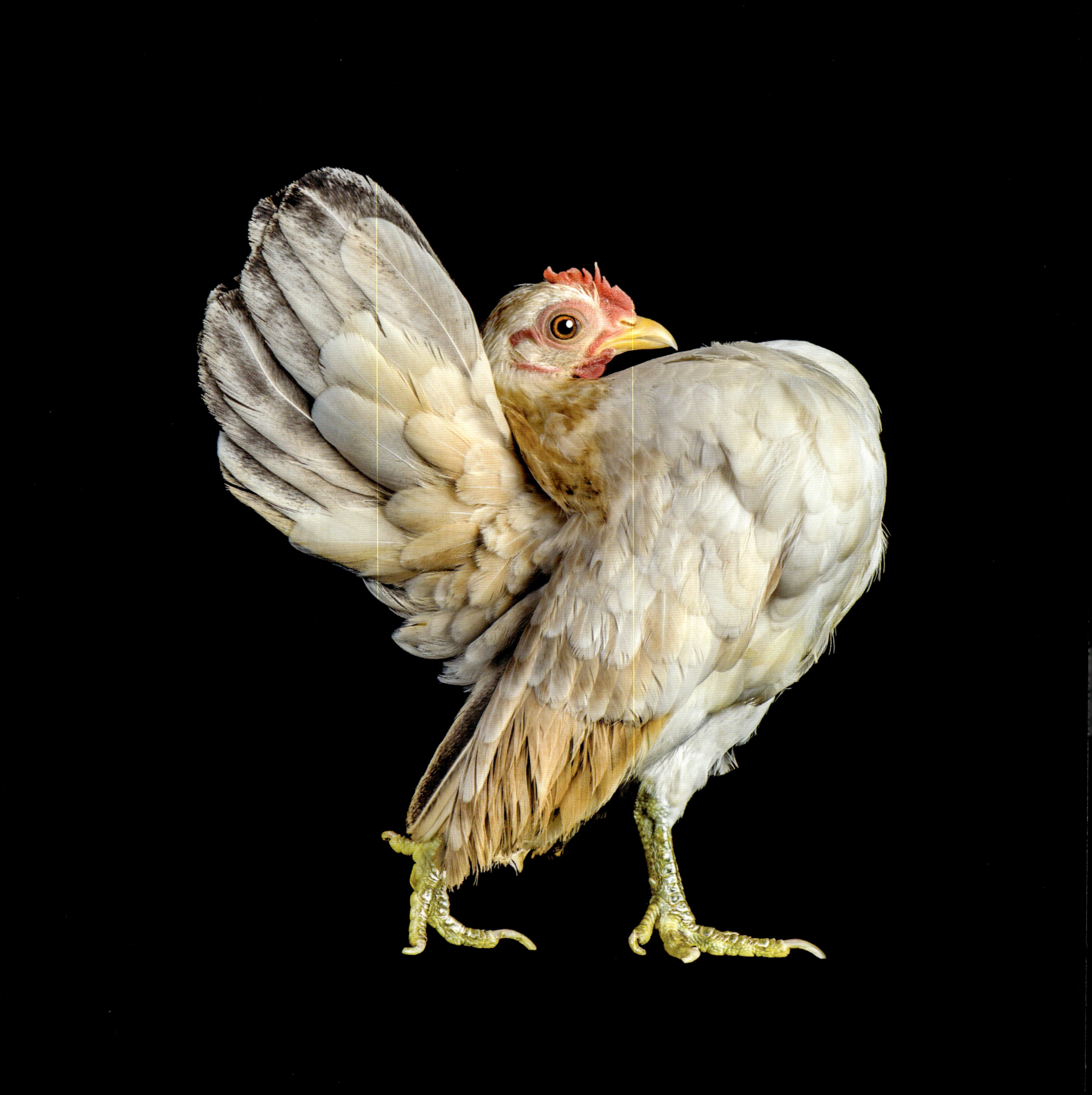

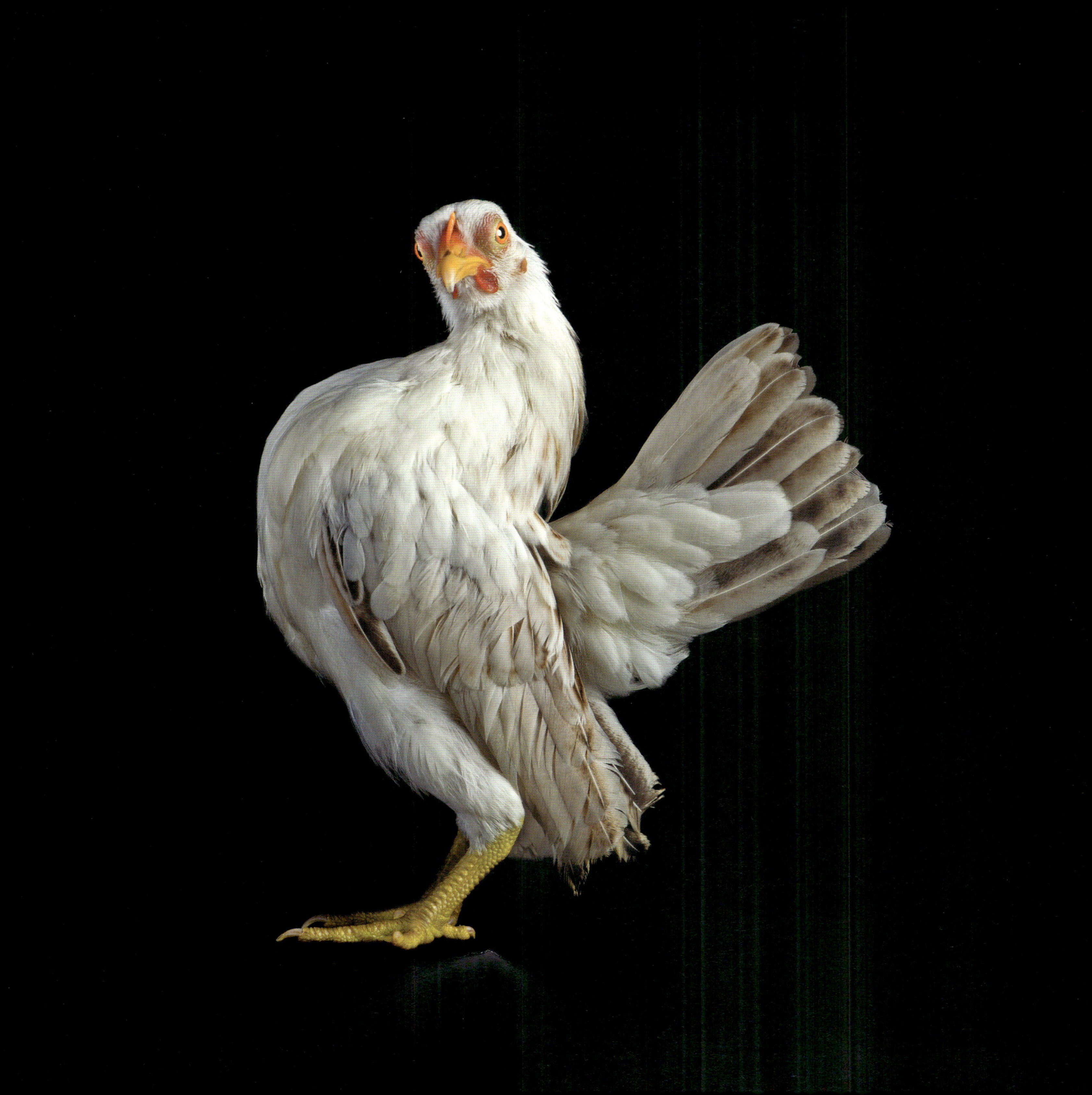

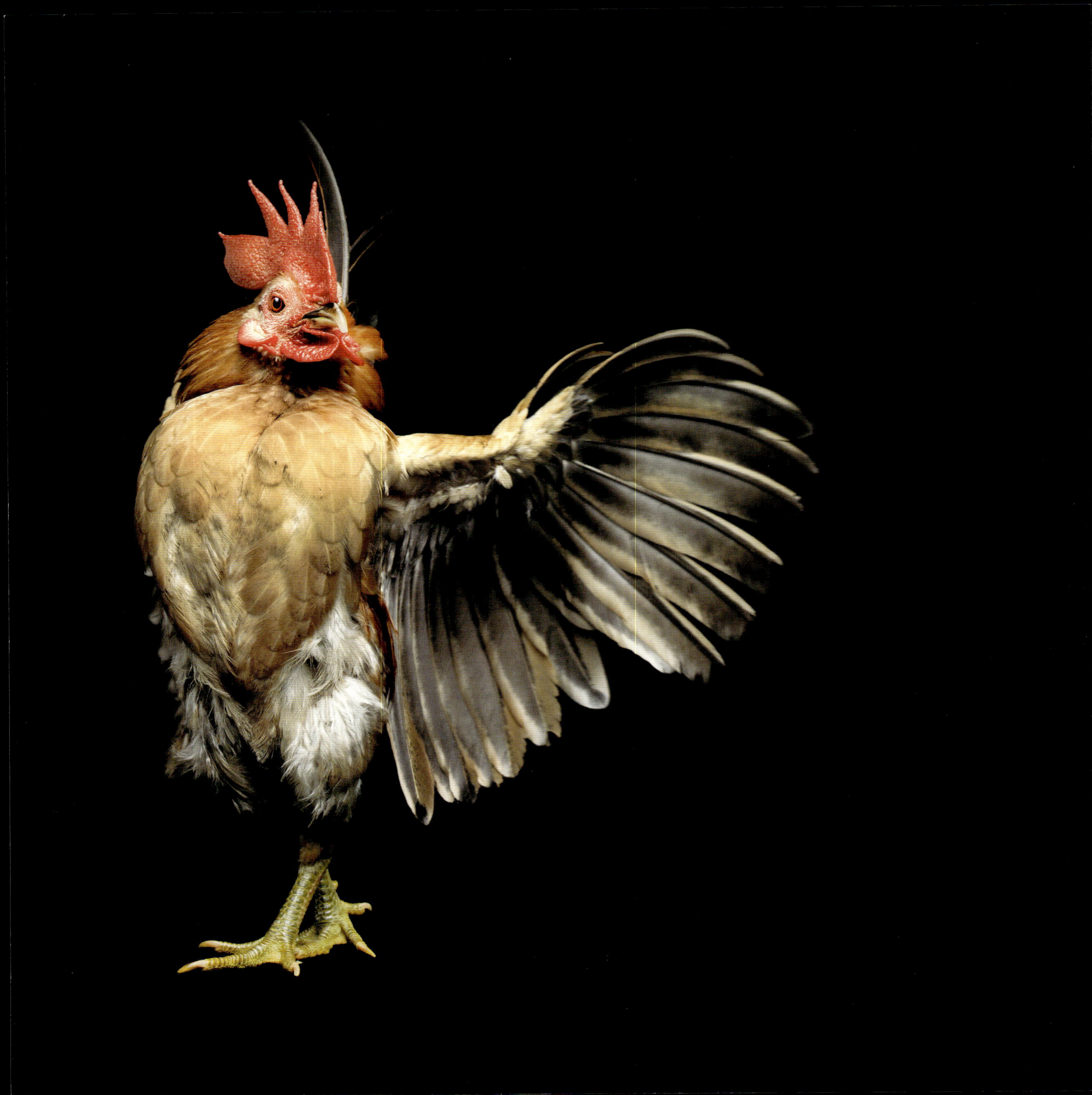

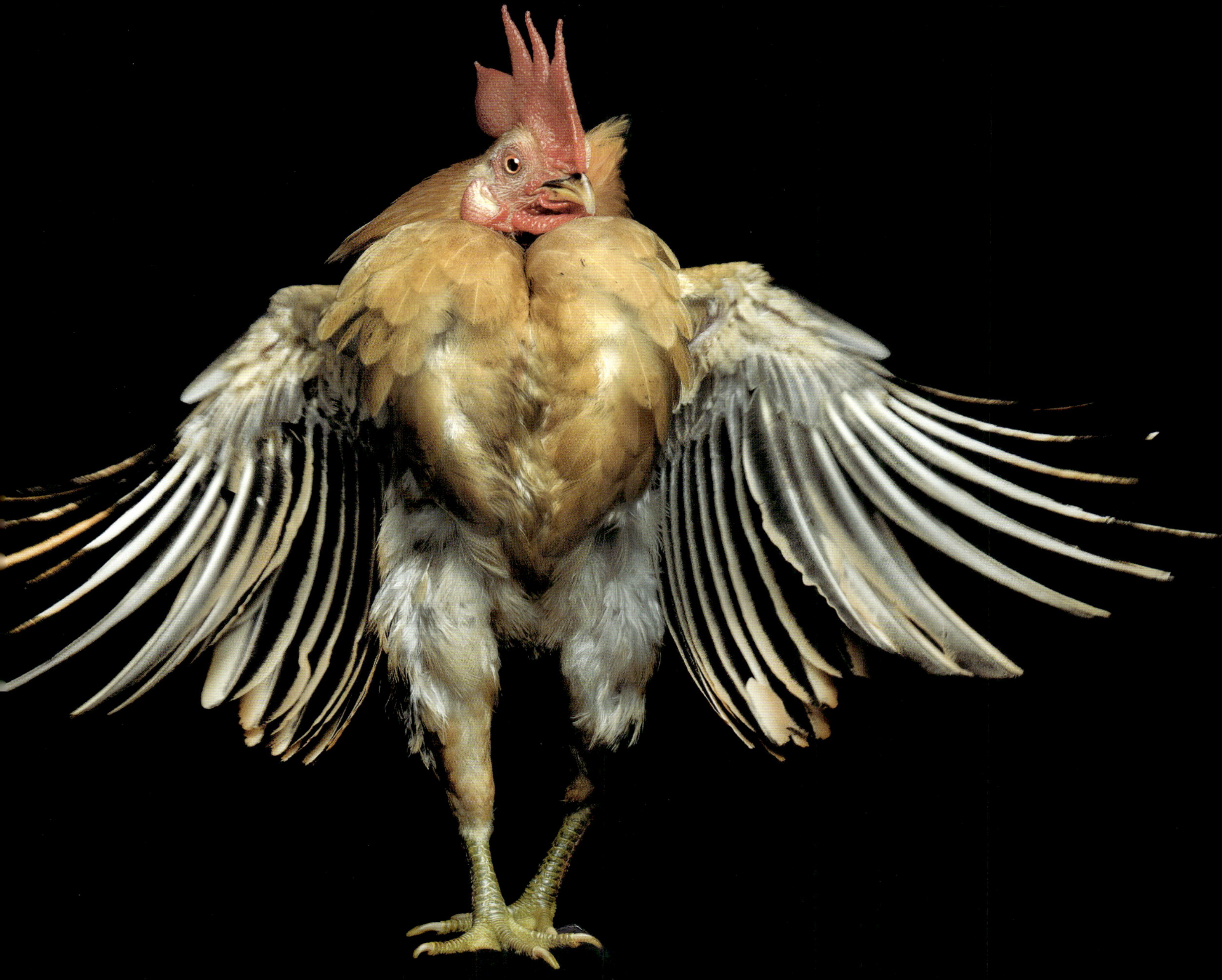

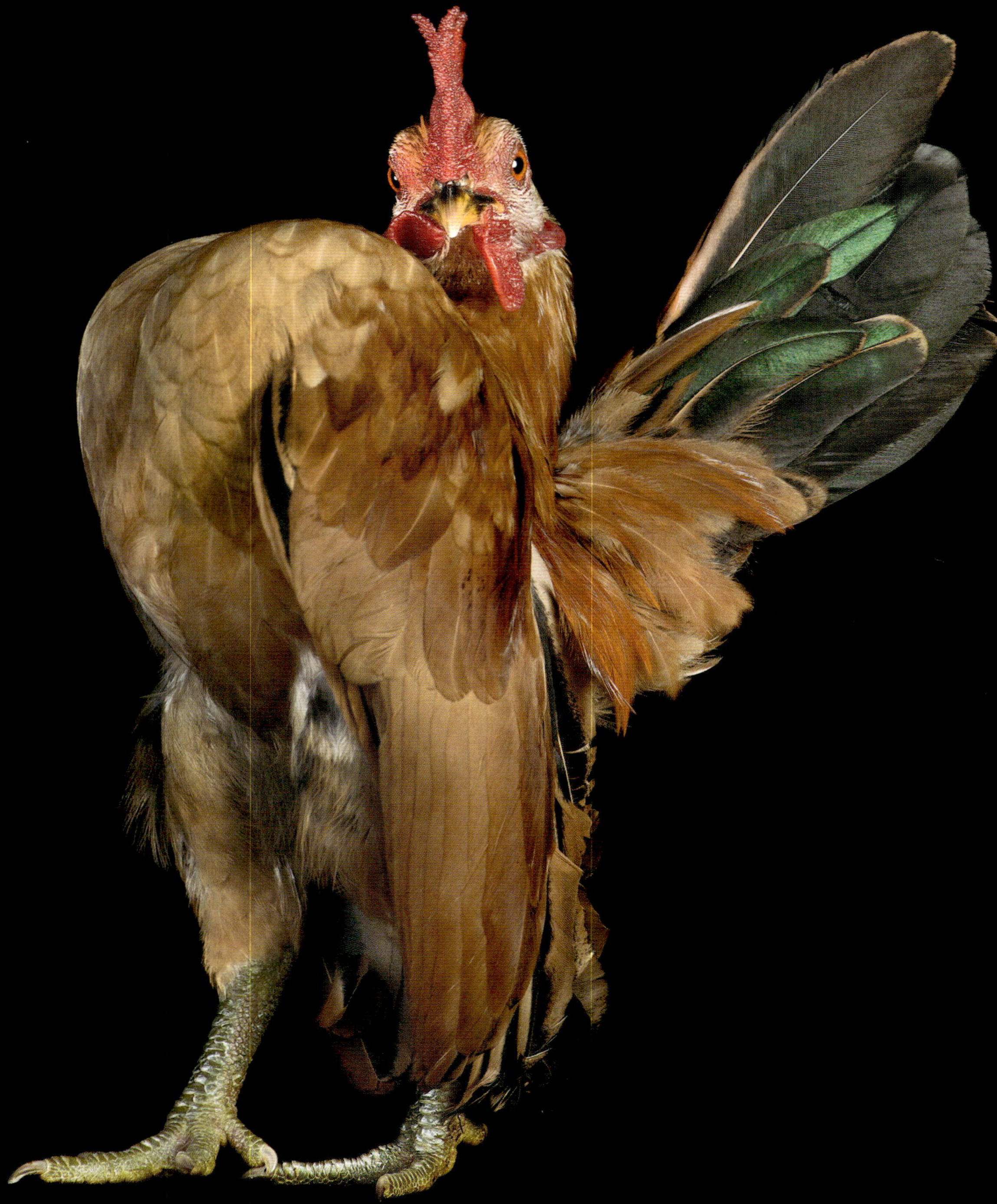

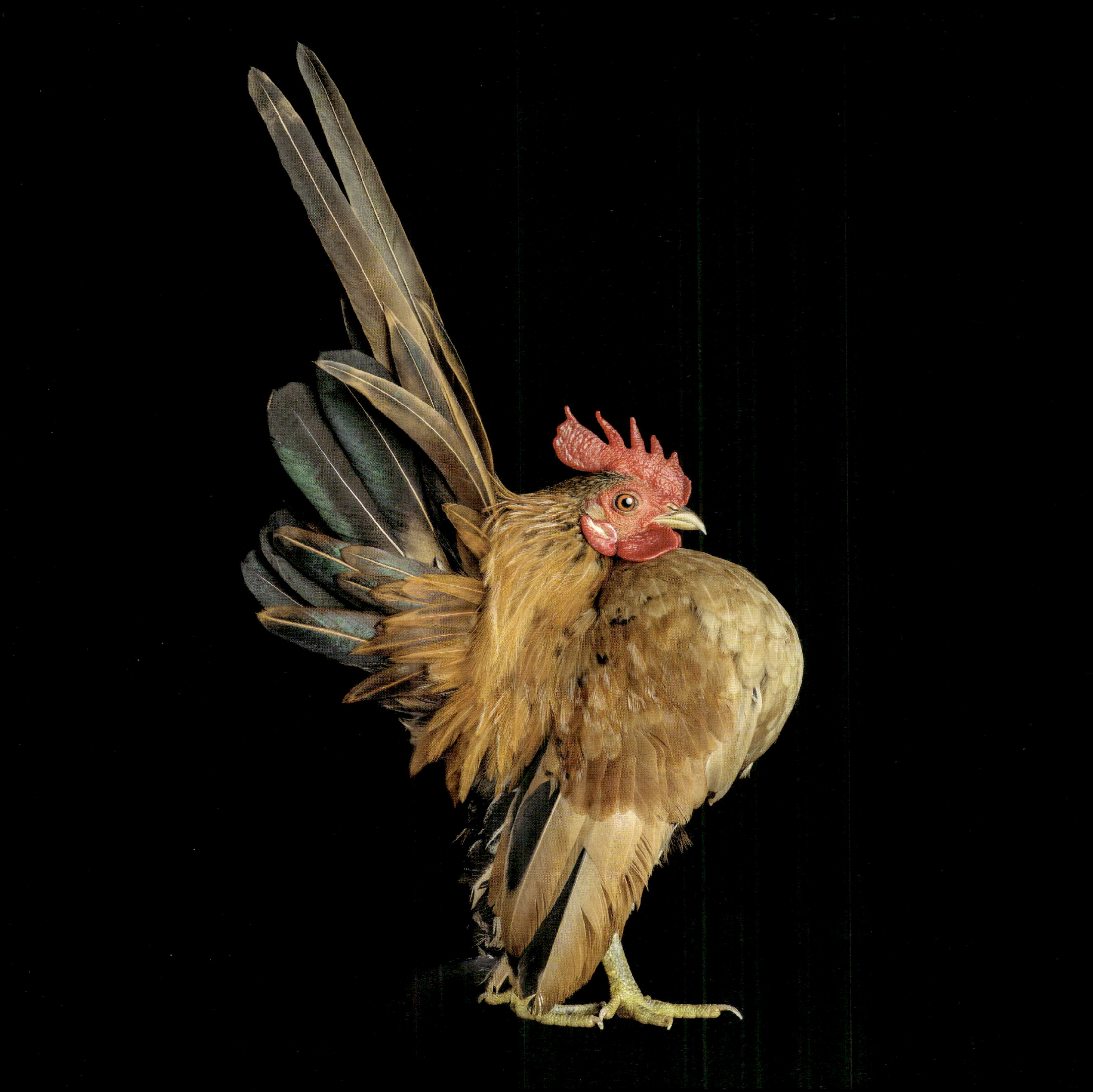

YOU CAN BE A ROOSTER ONE DAY,

AND A FEATHER DUSTER THE NEXT.

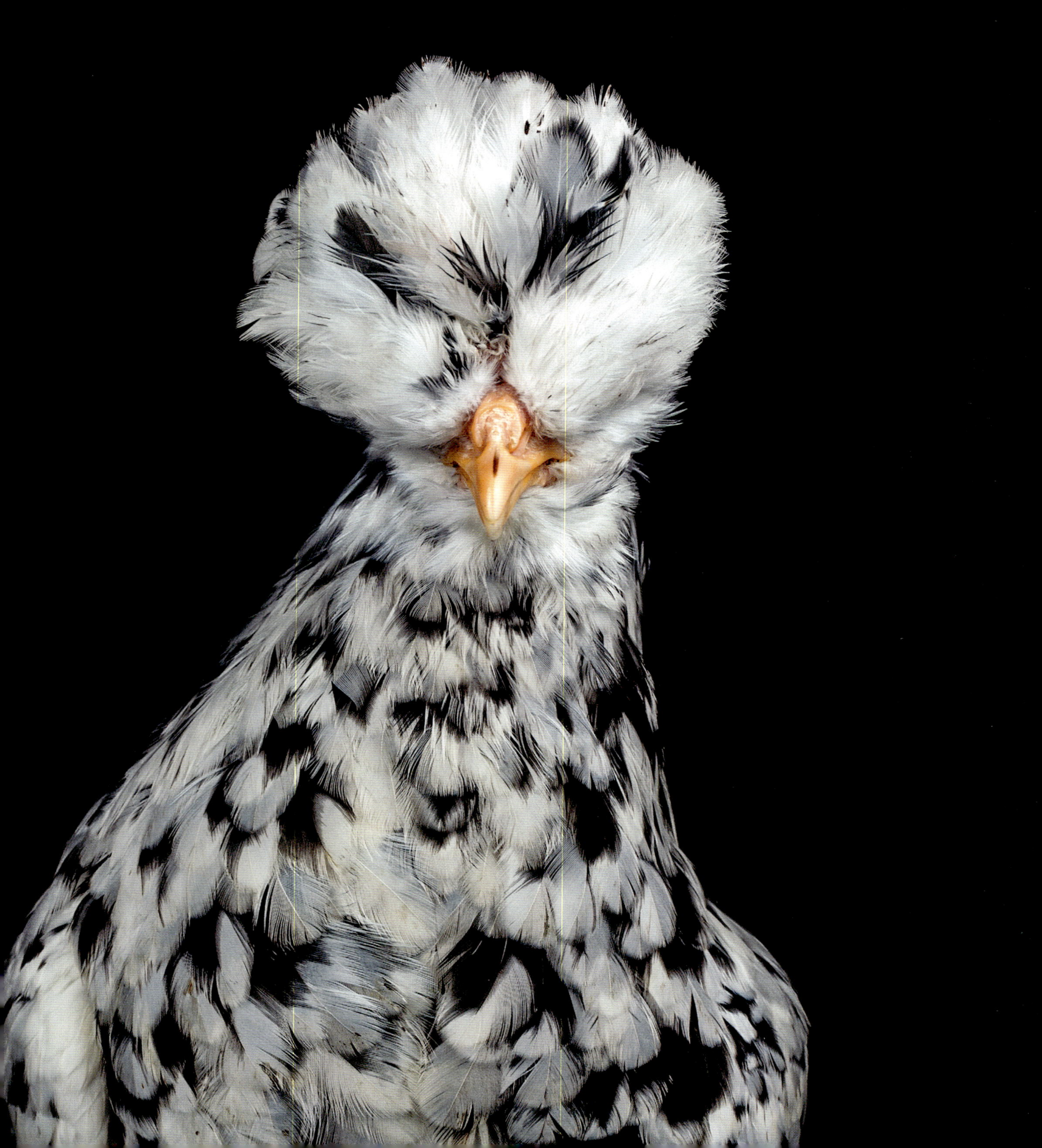

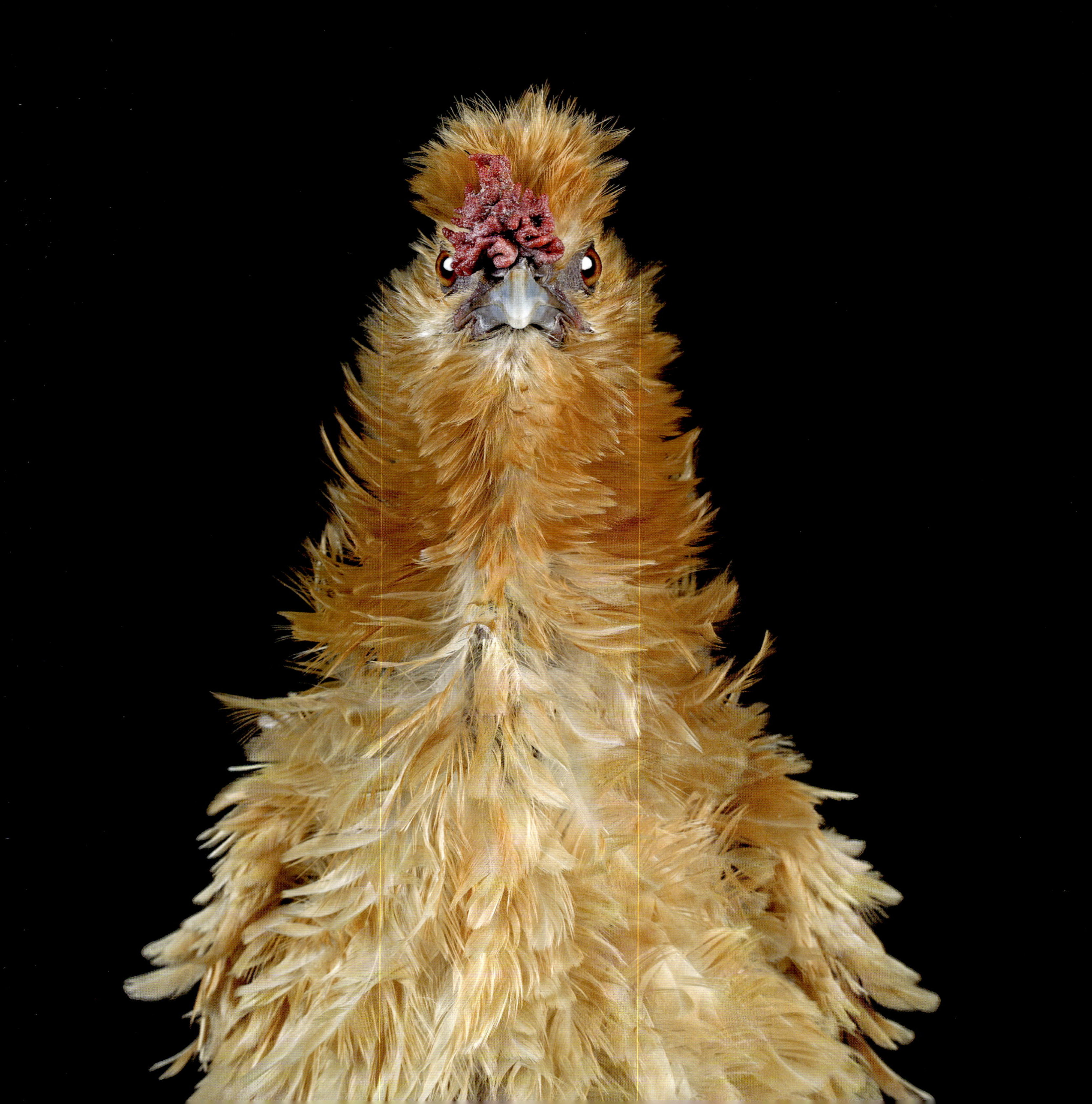

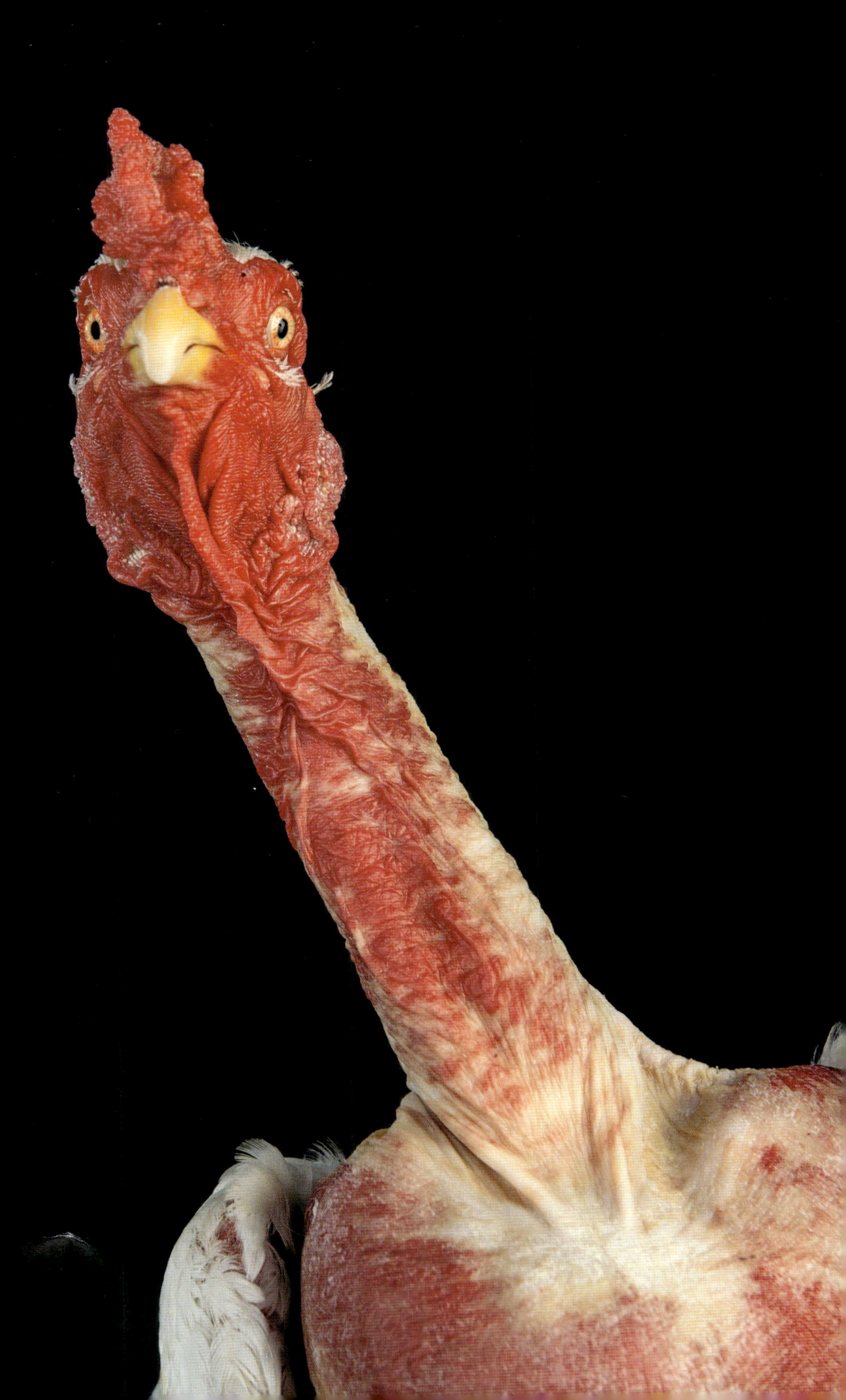

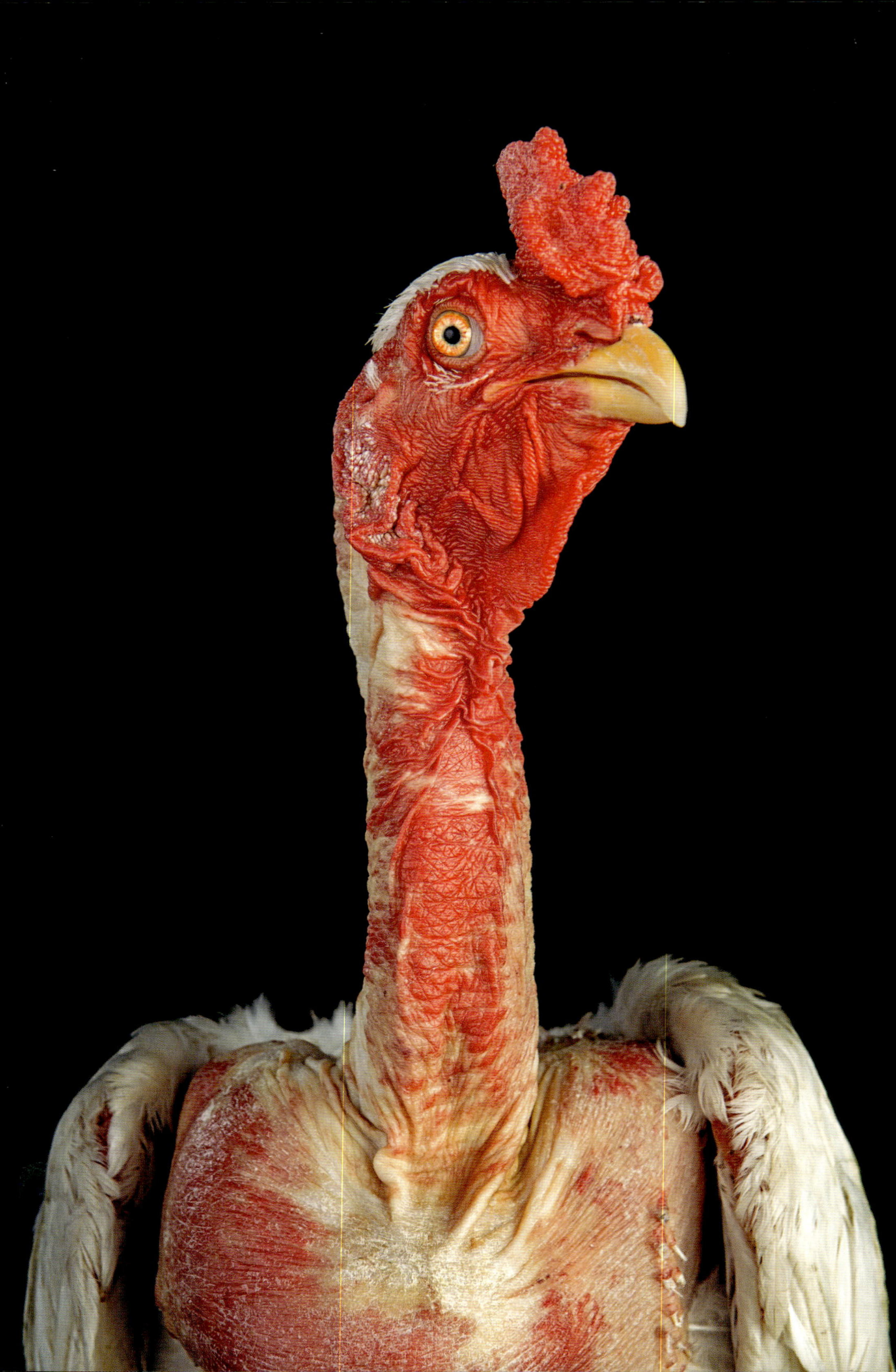

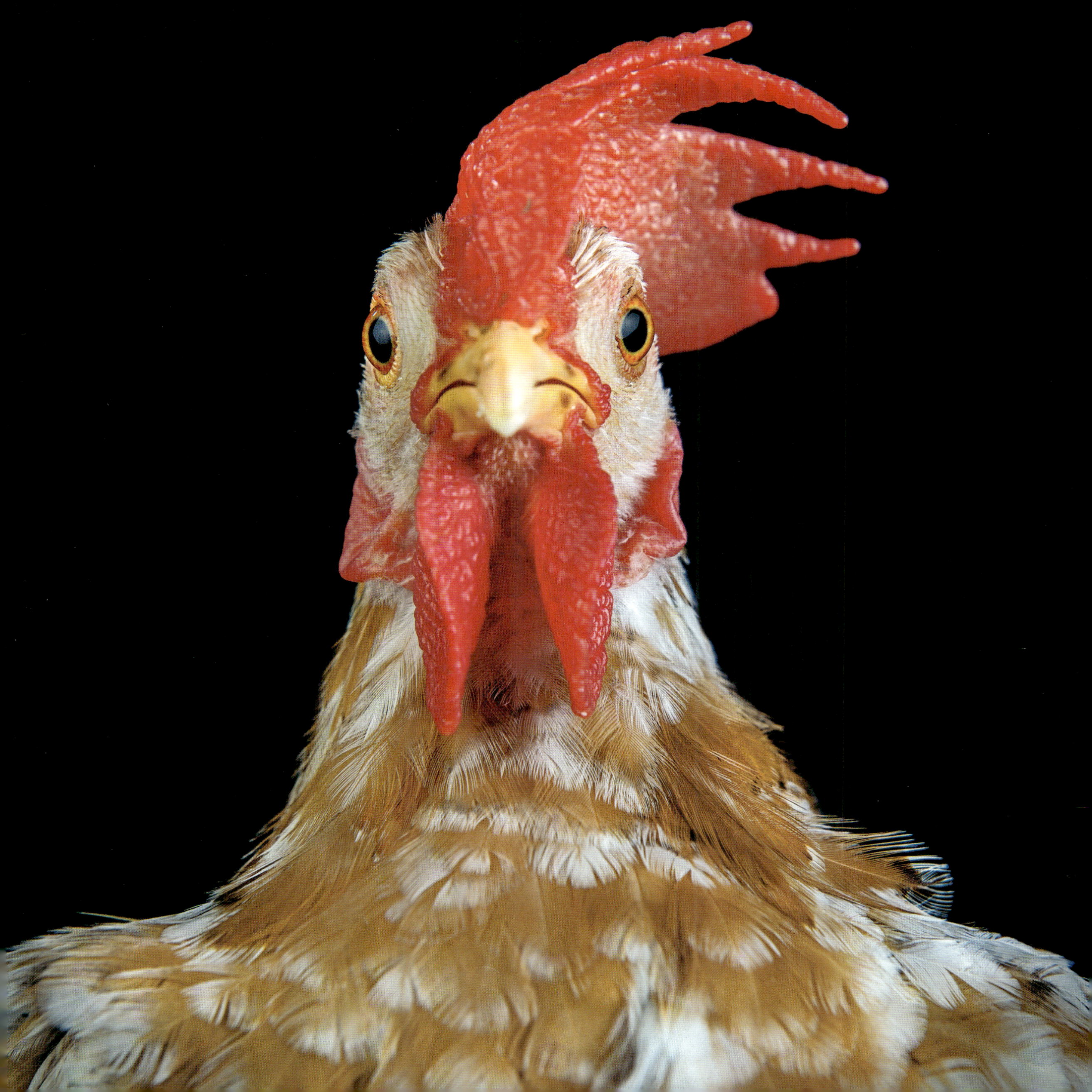

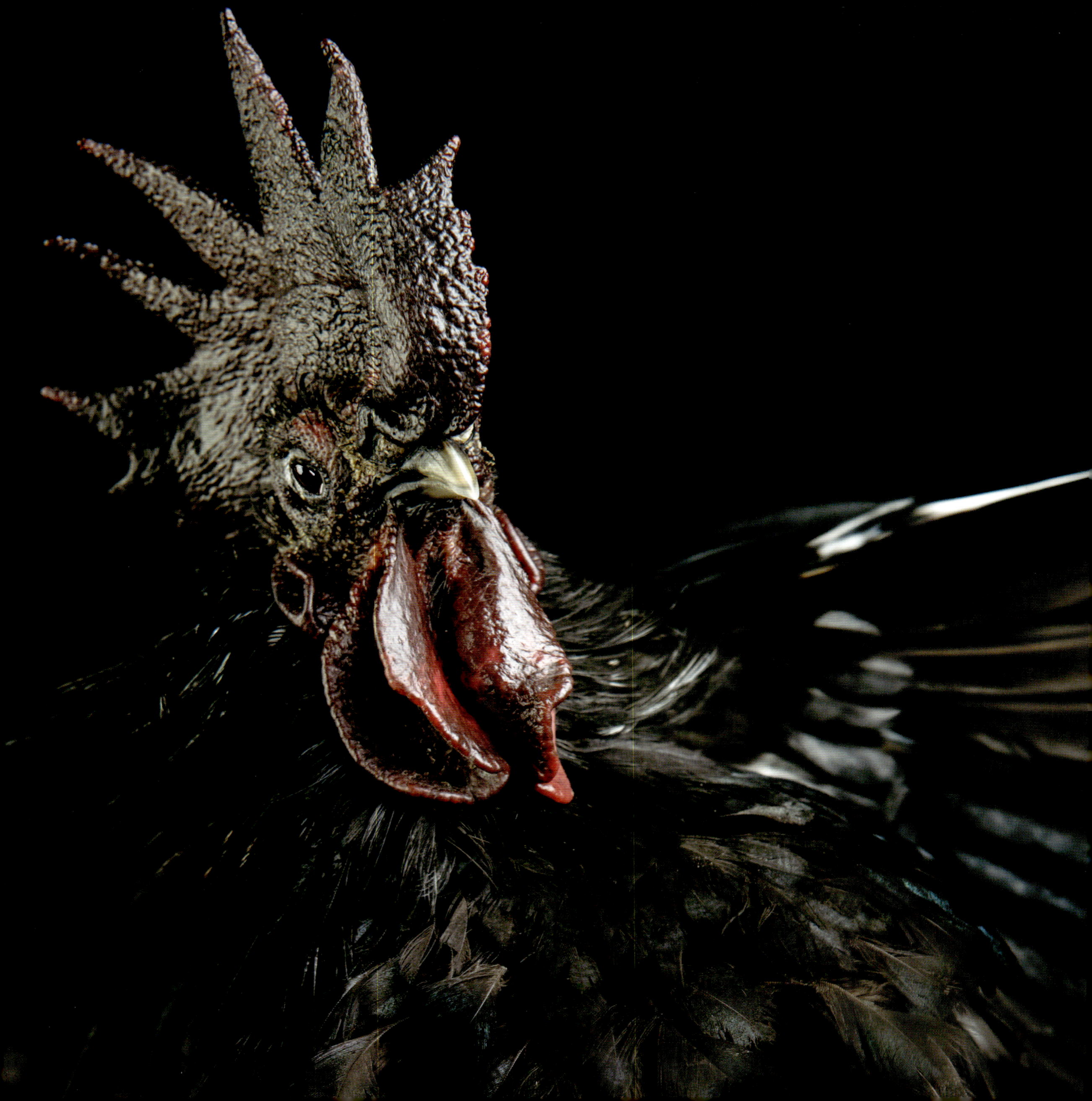

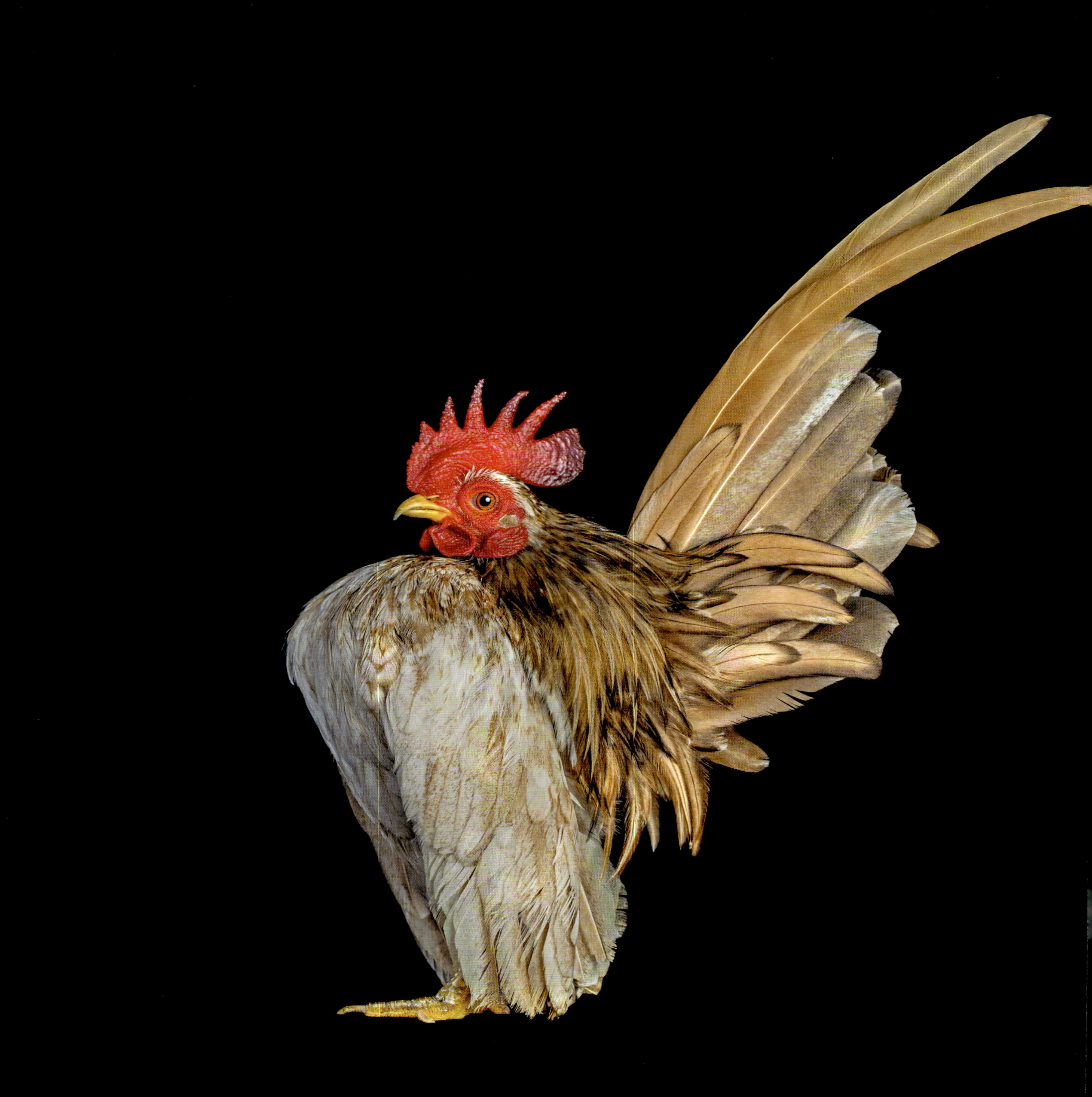

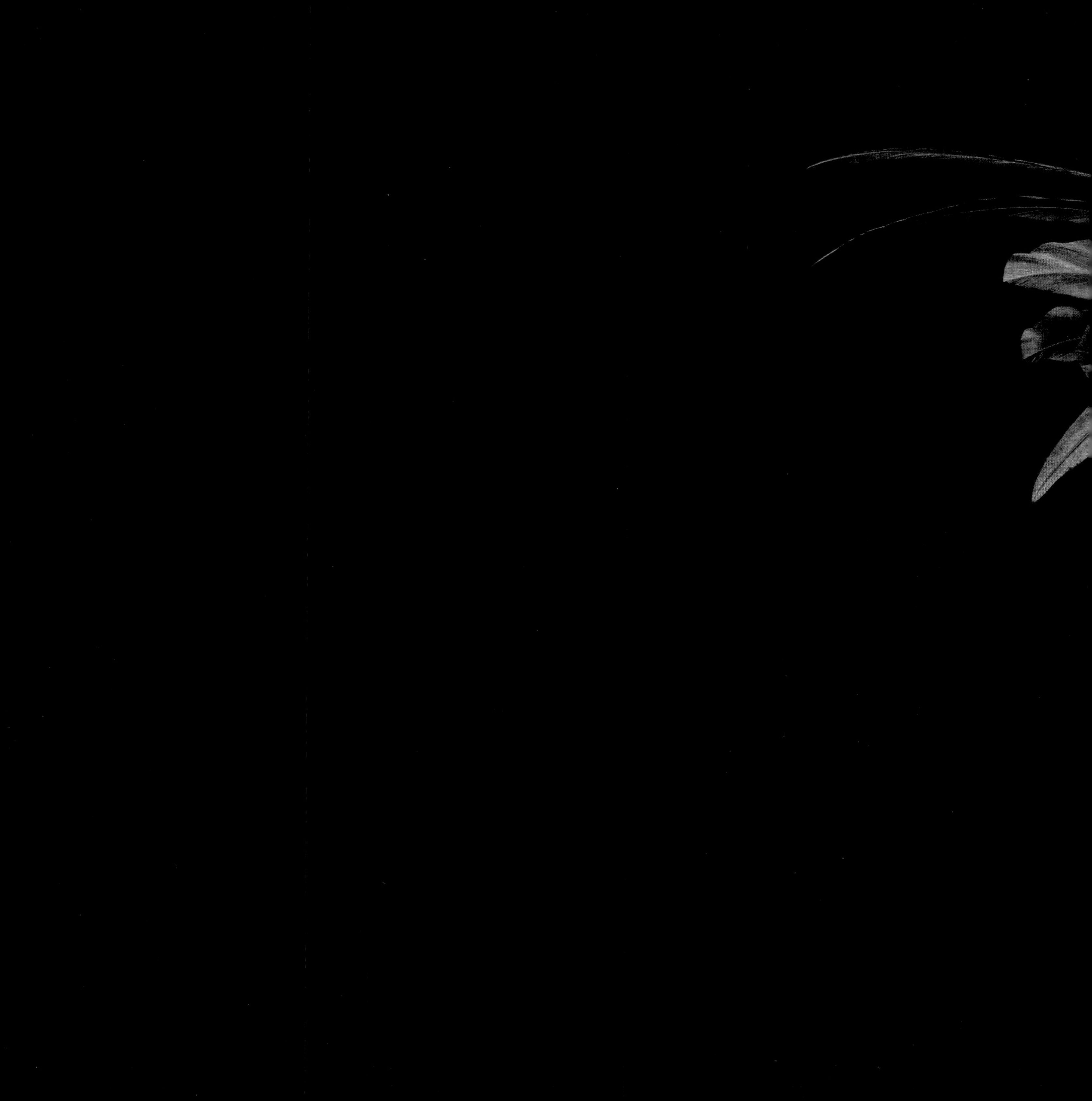

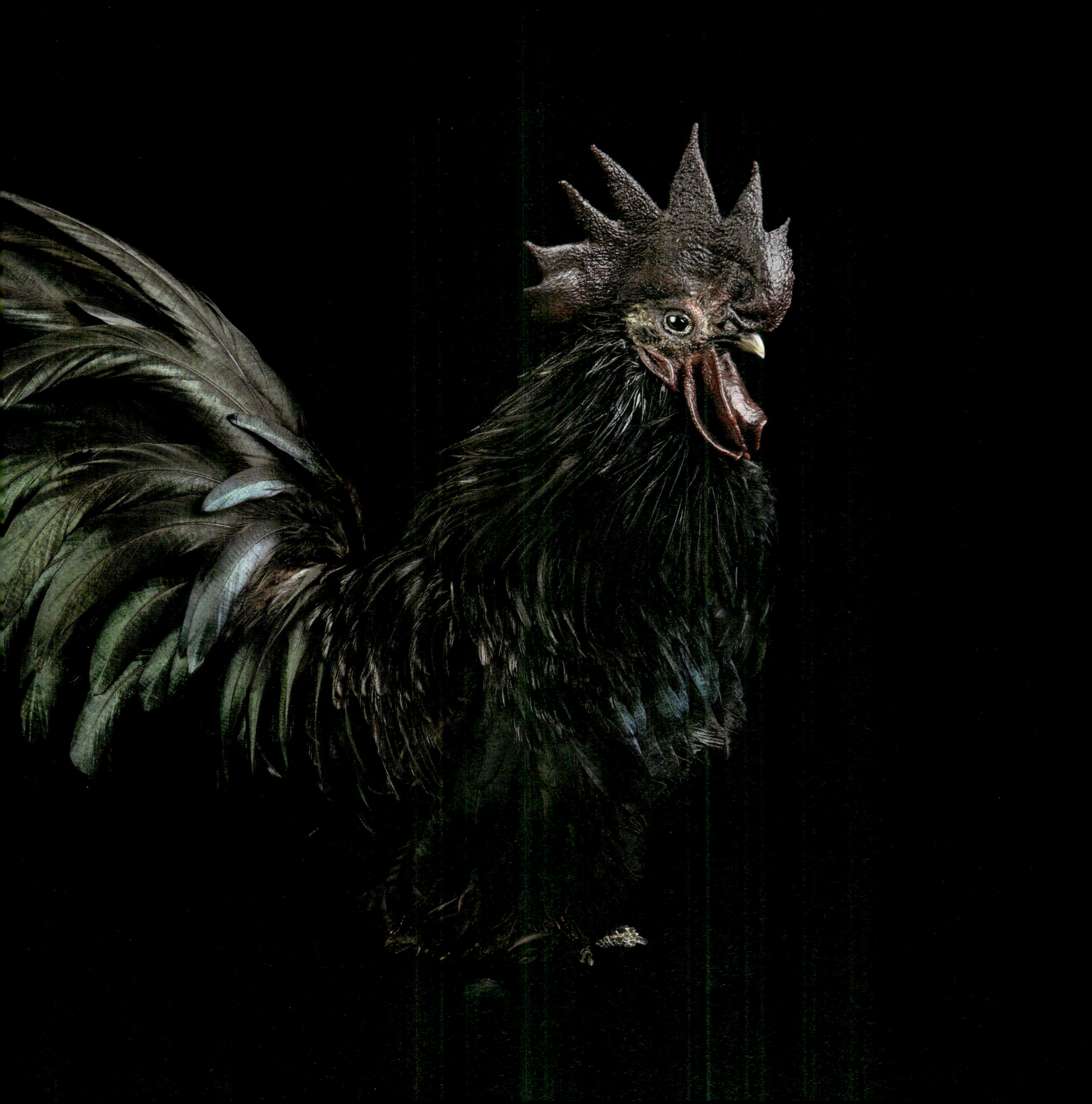

HEN

PECKED

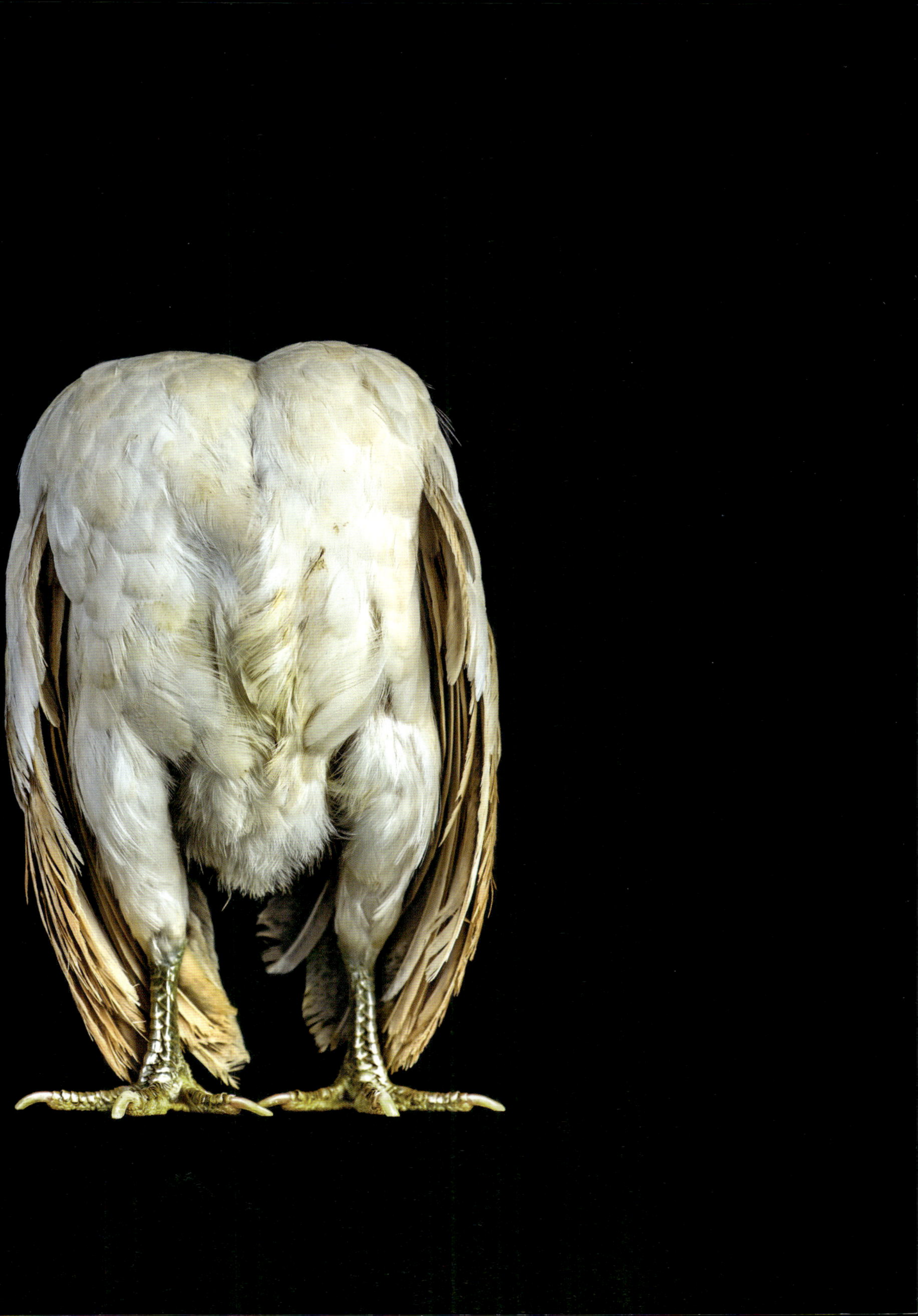

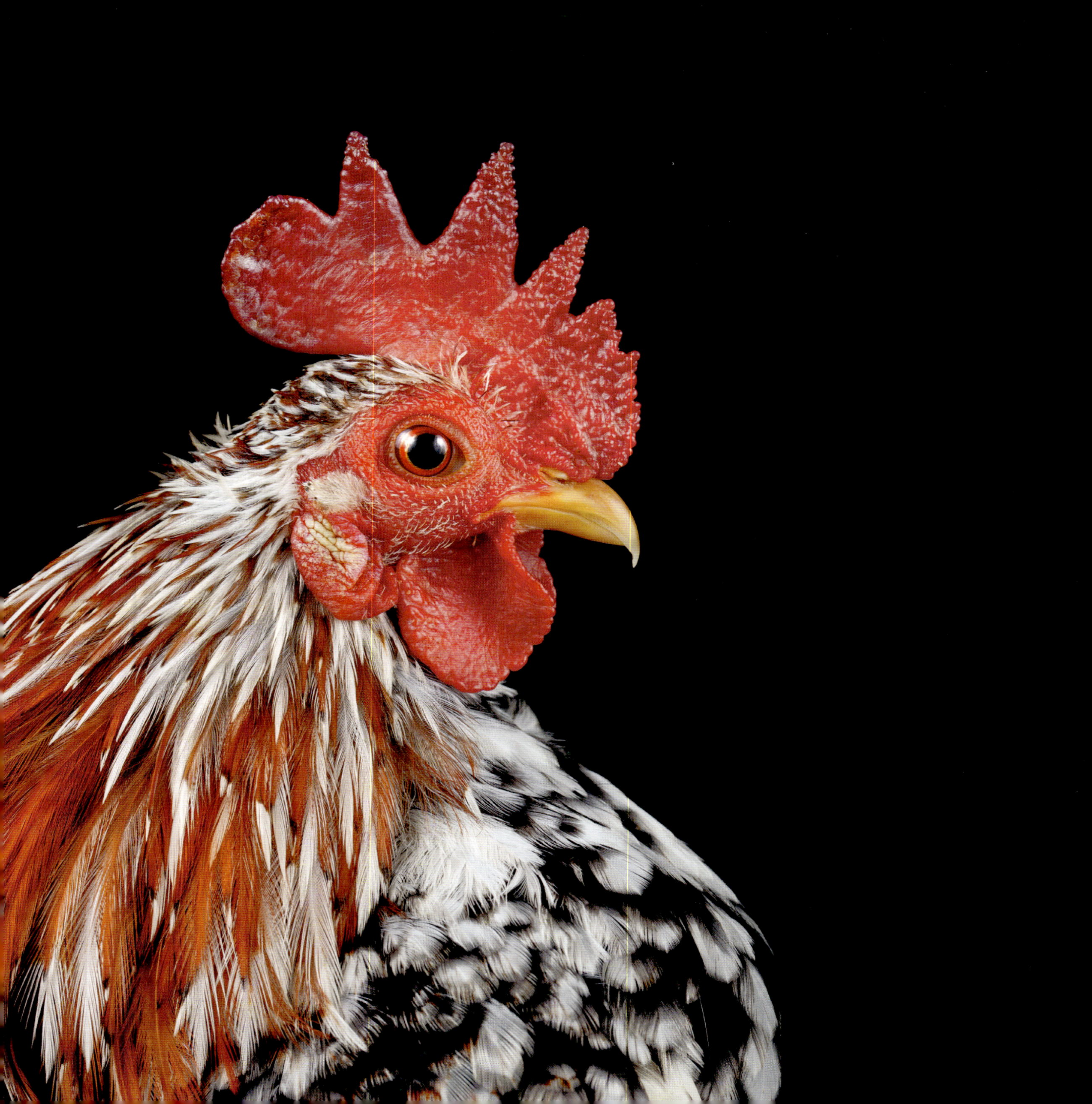

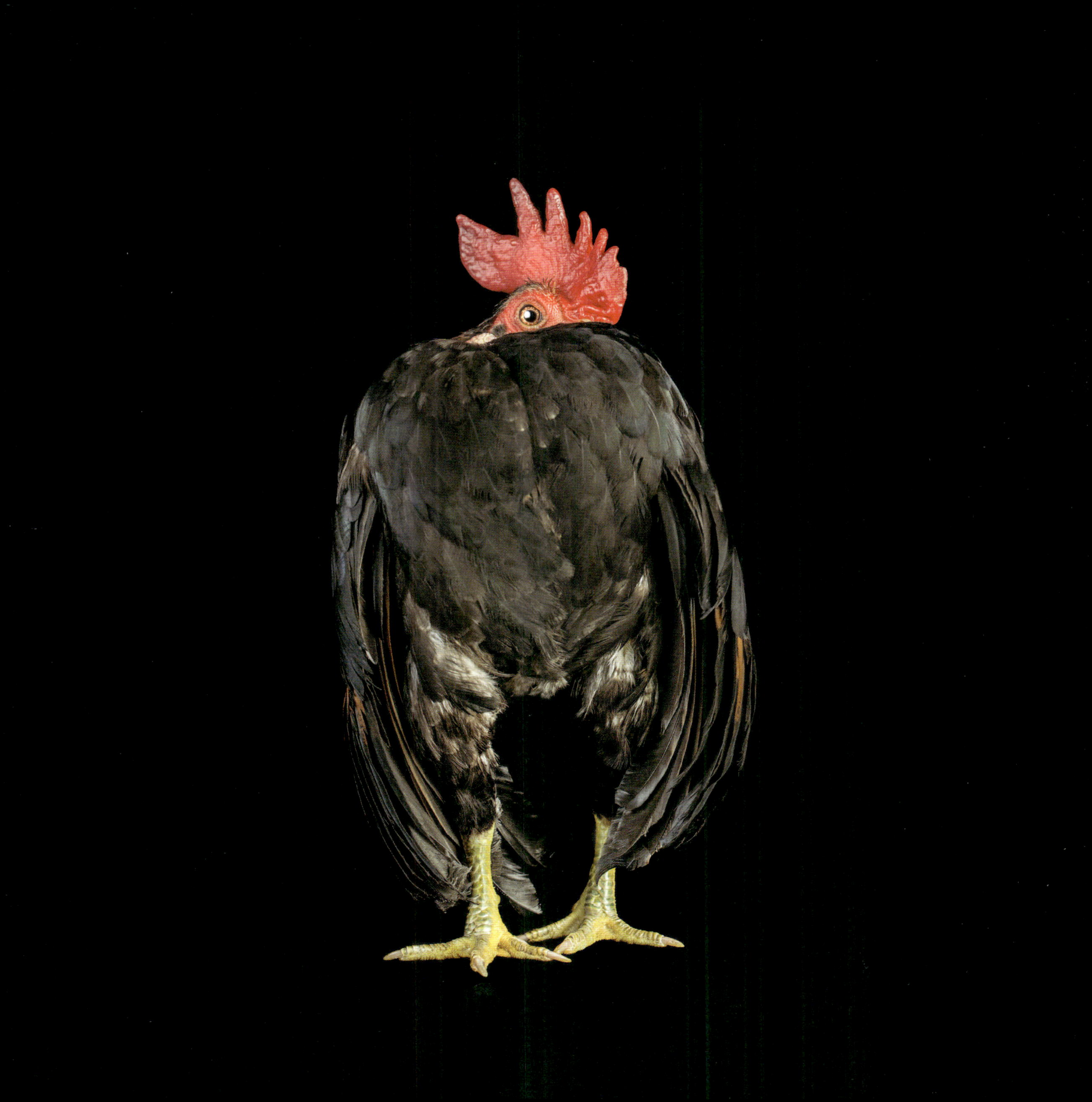

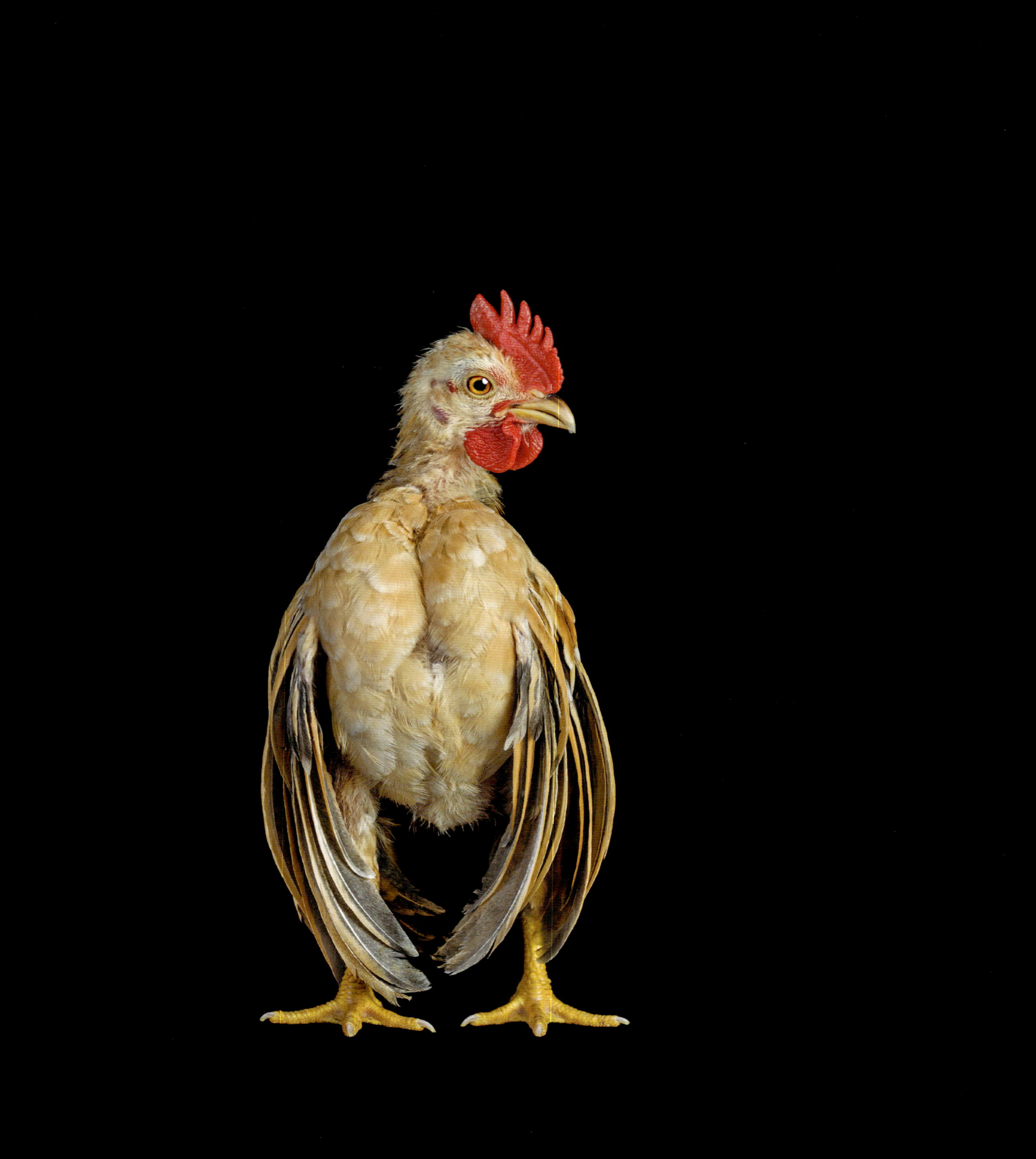

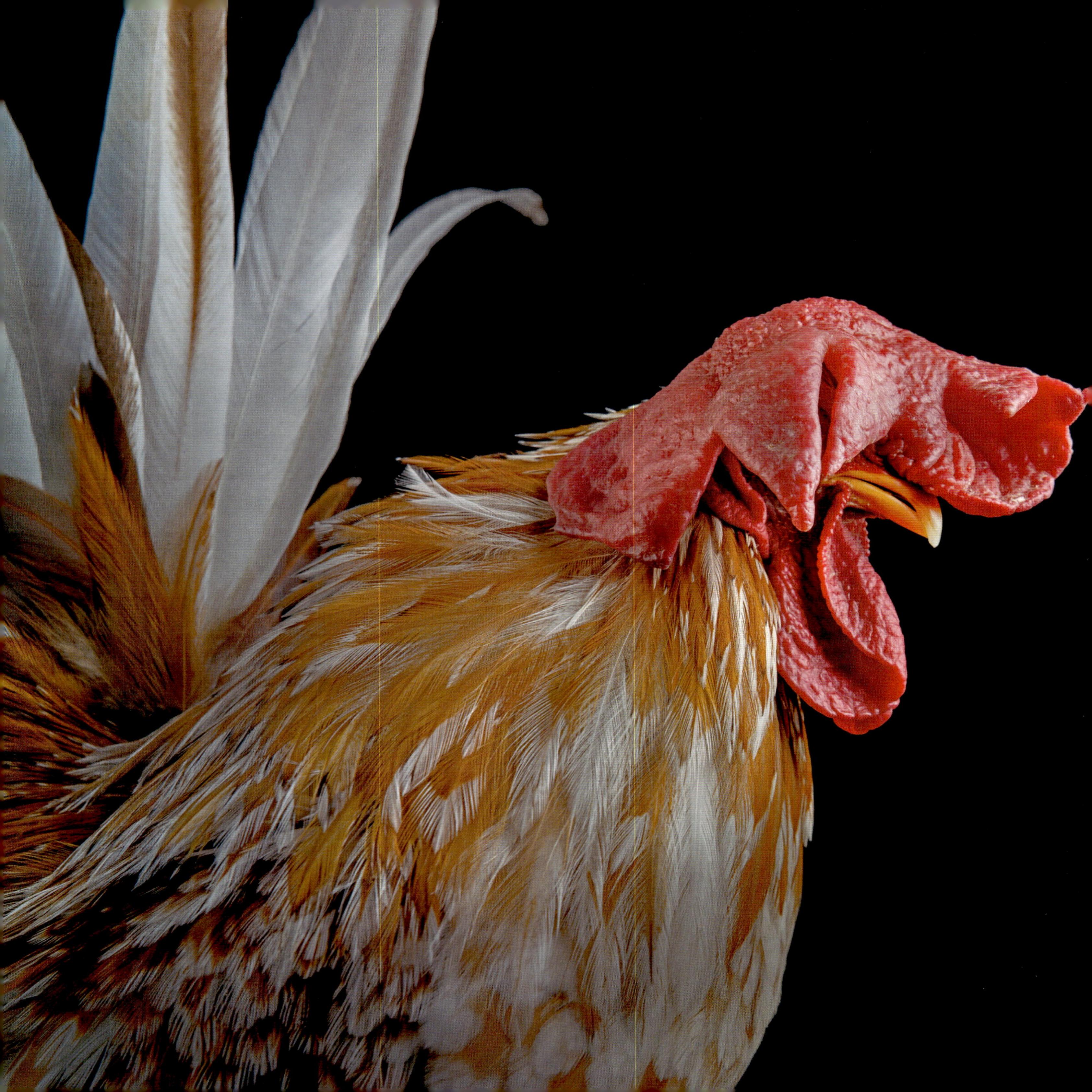

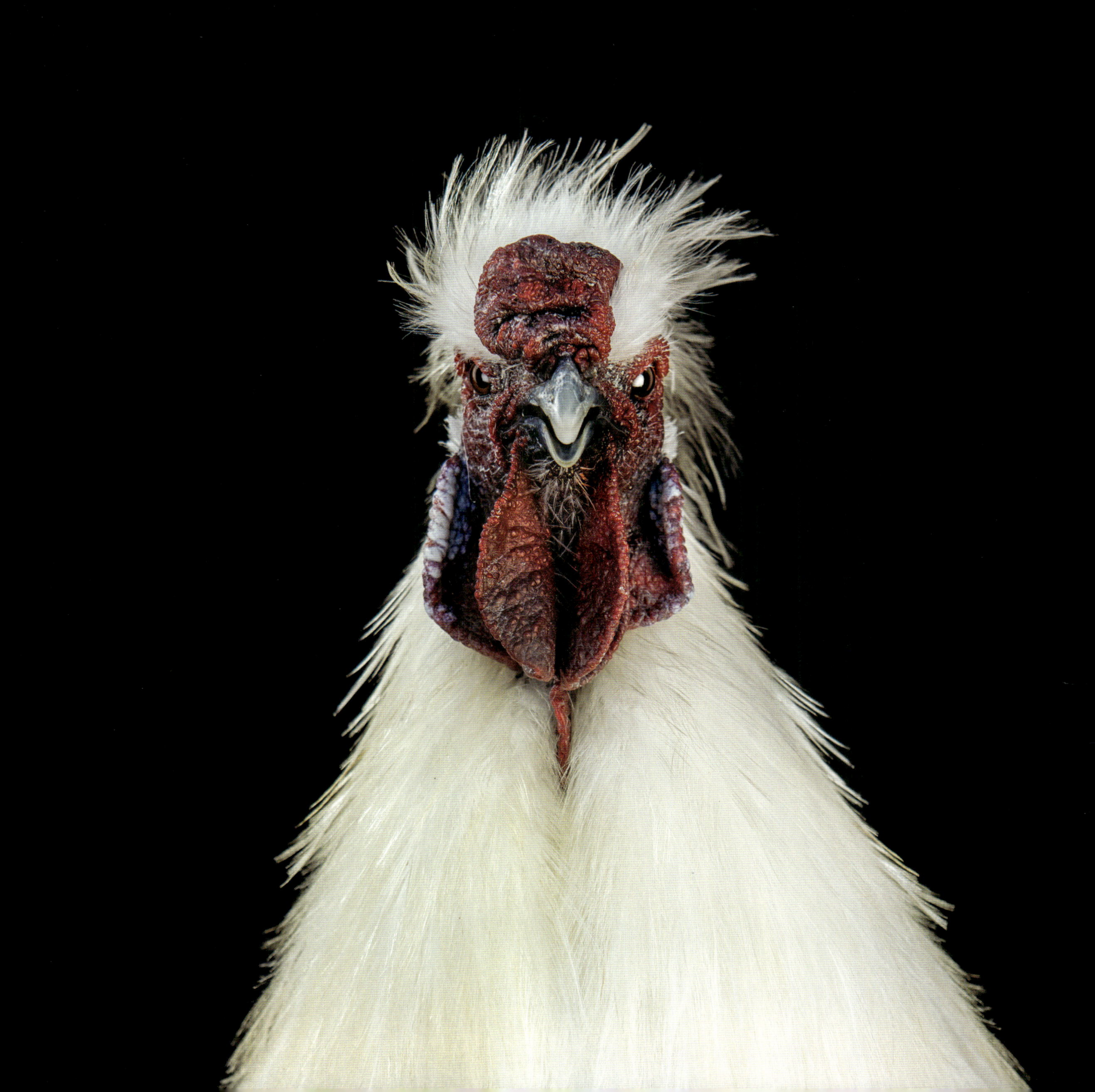

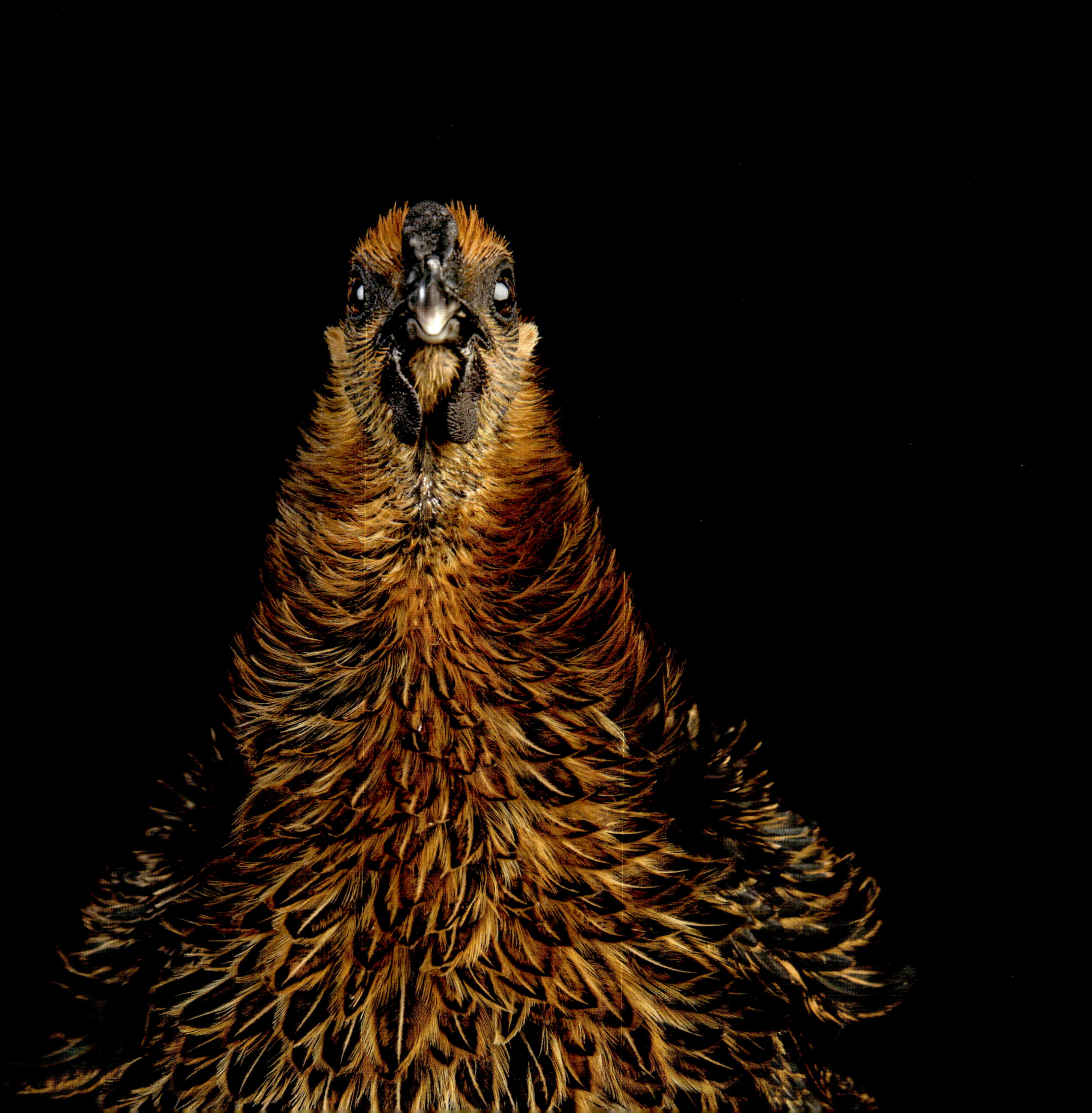

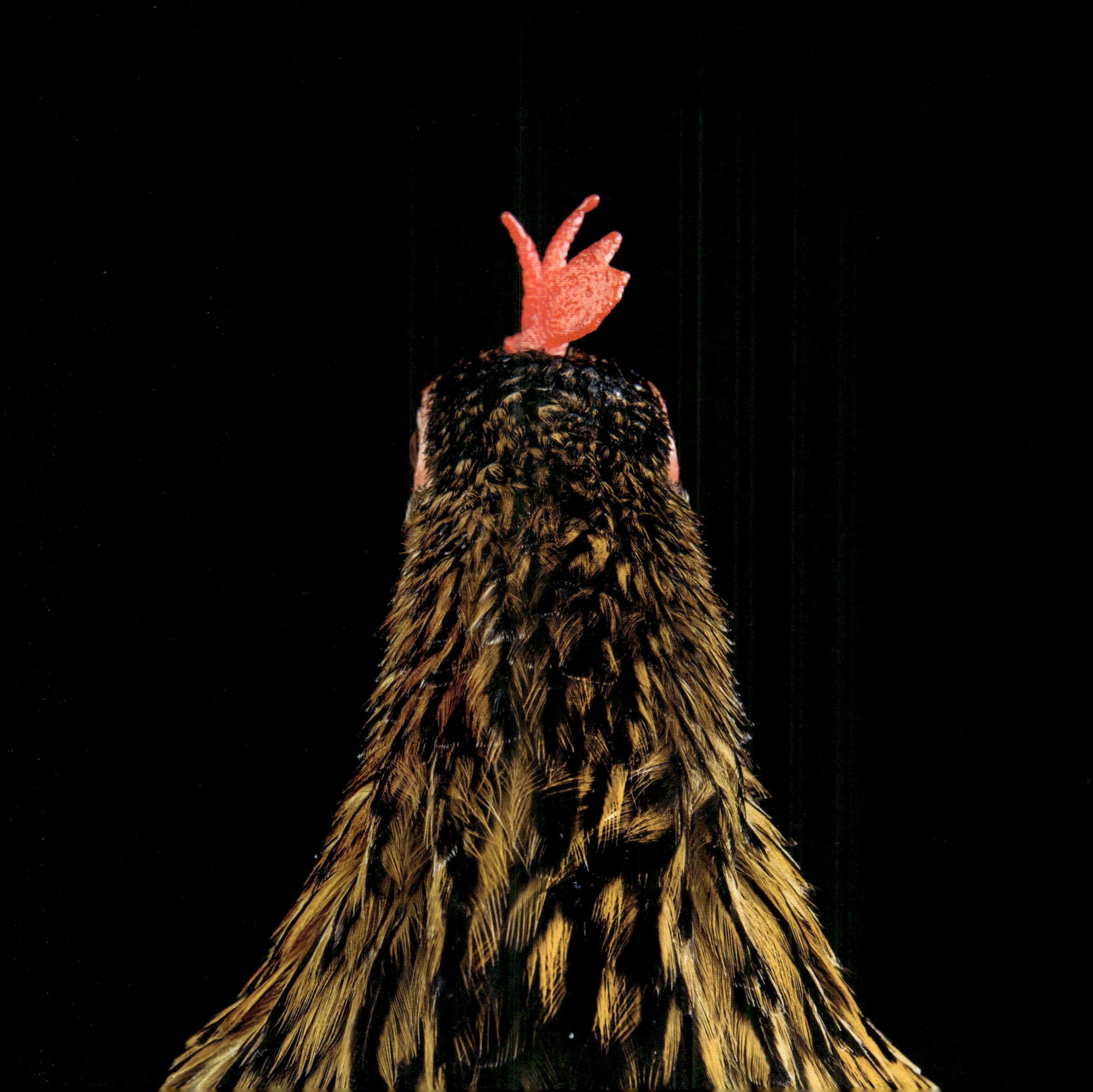

WHY DID THE CHICKEN

CROSS THE ROAD?

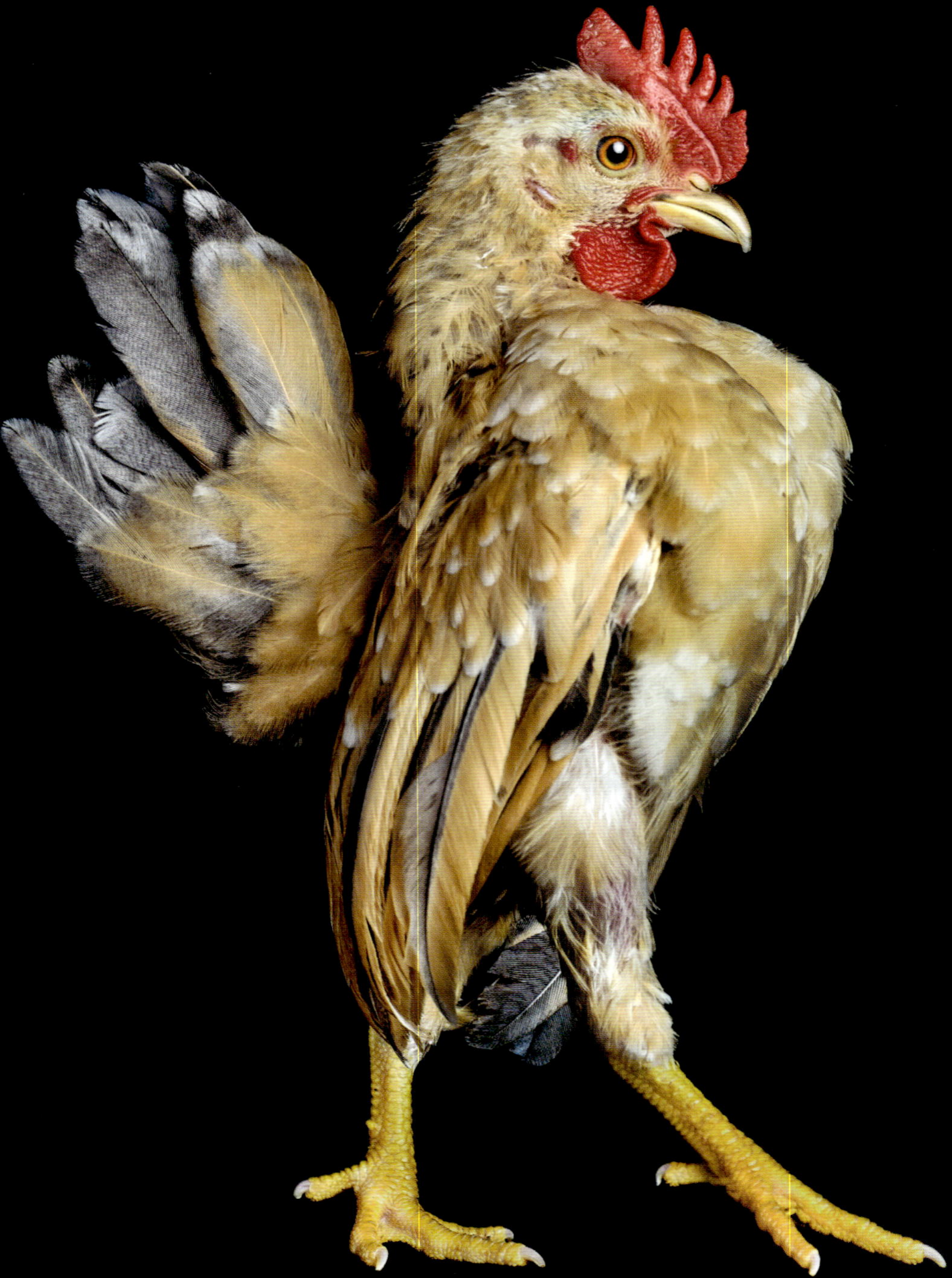

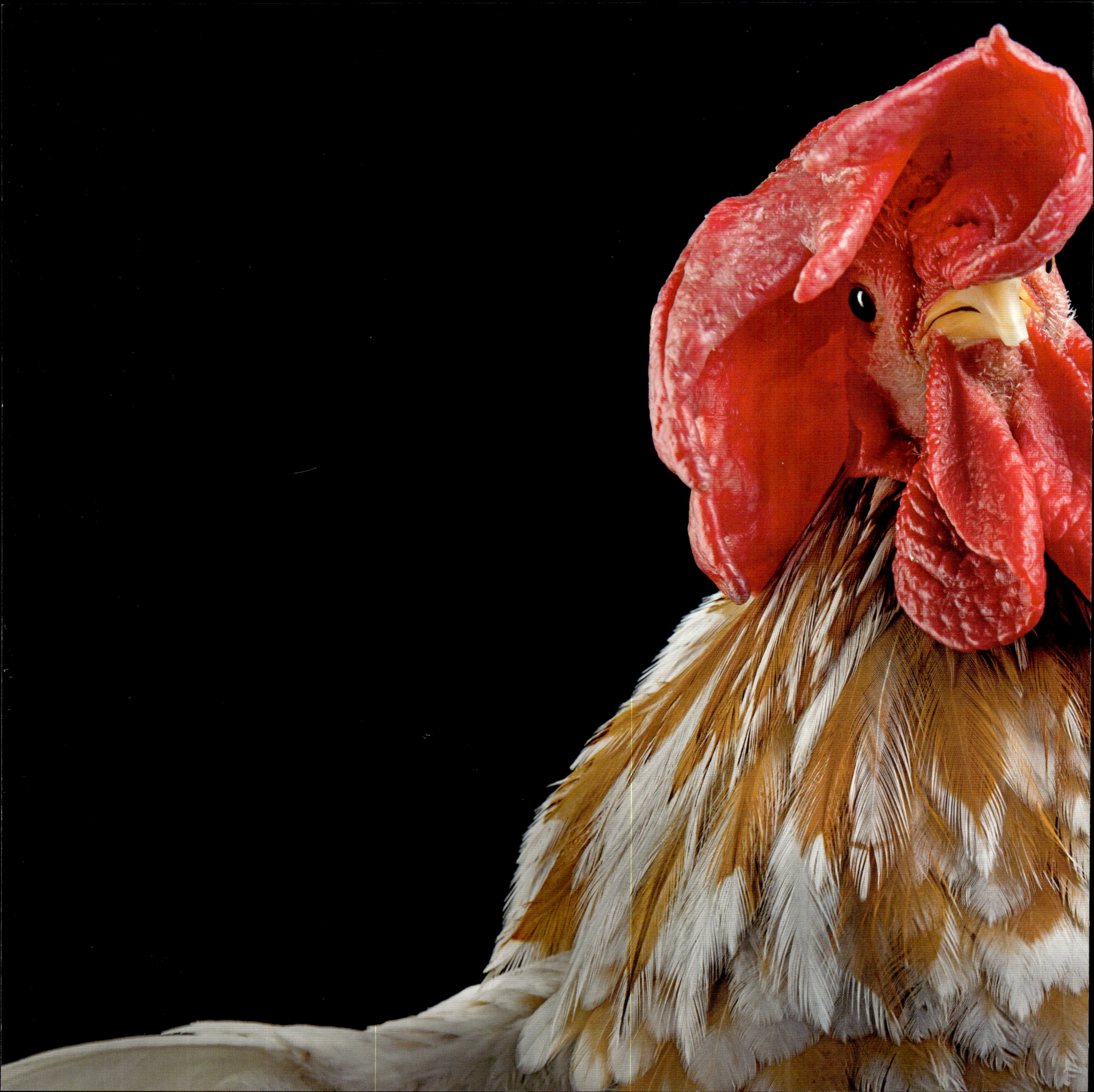

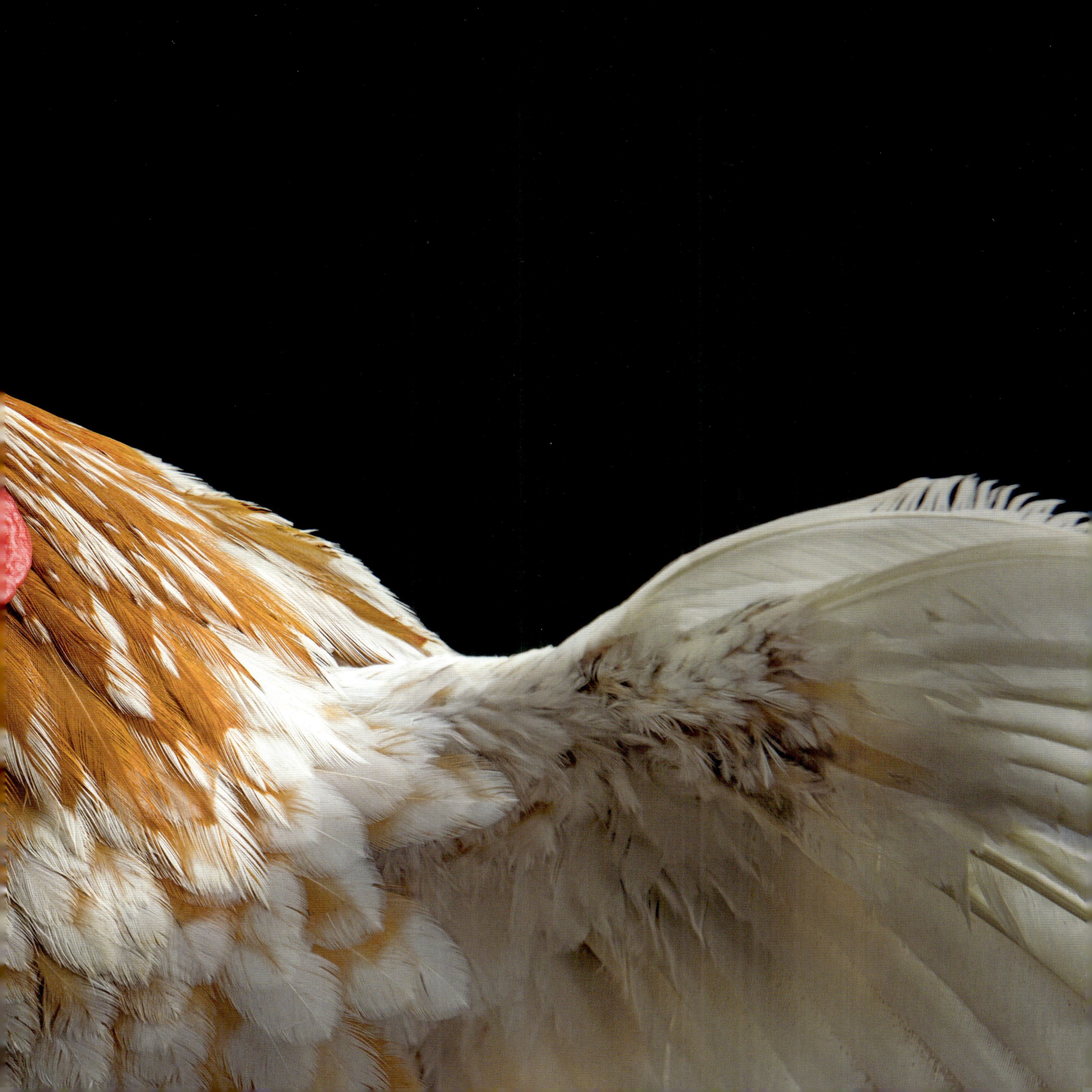

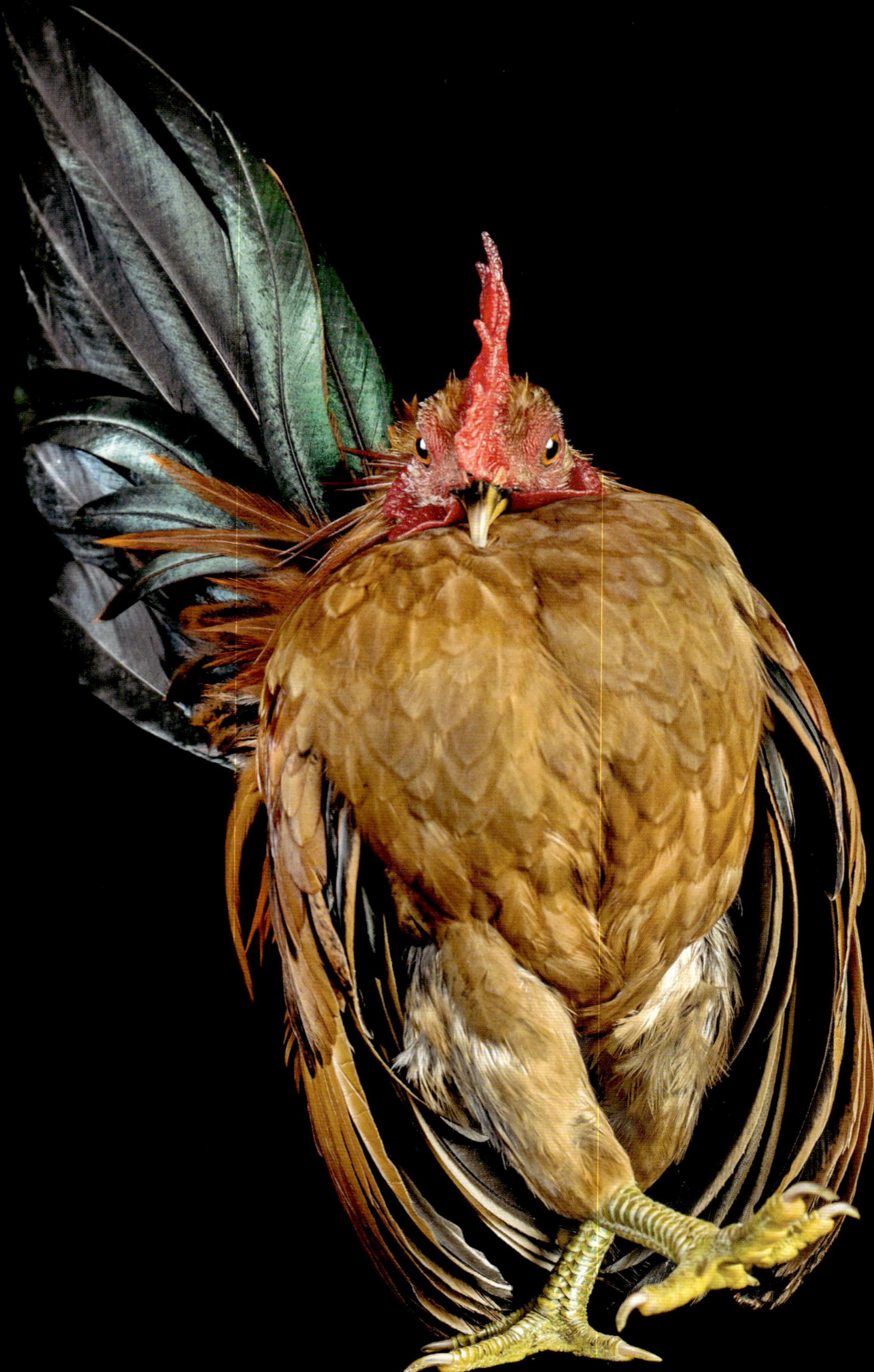

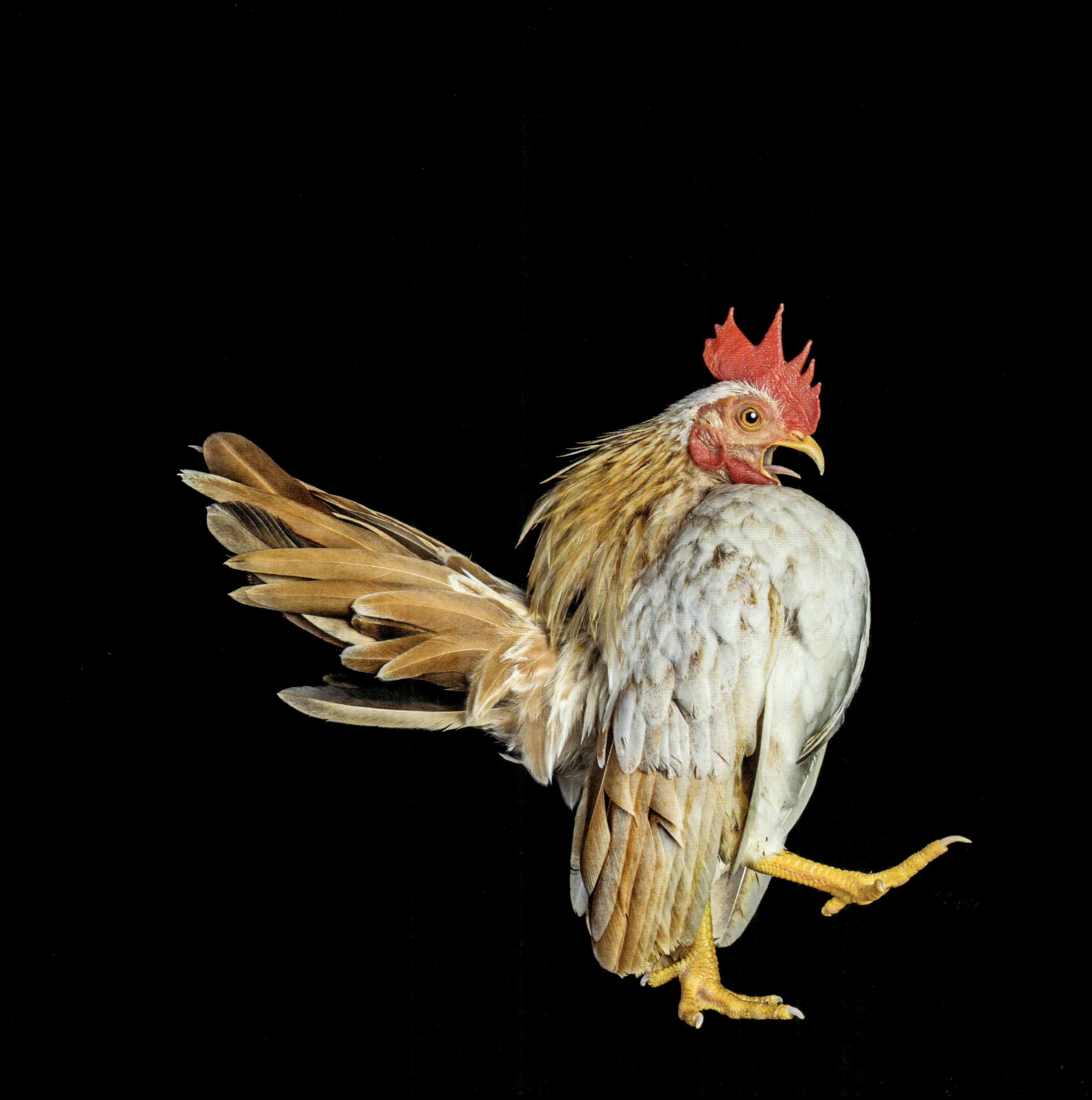

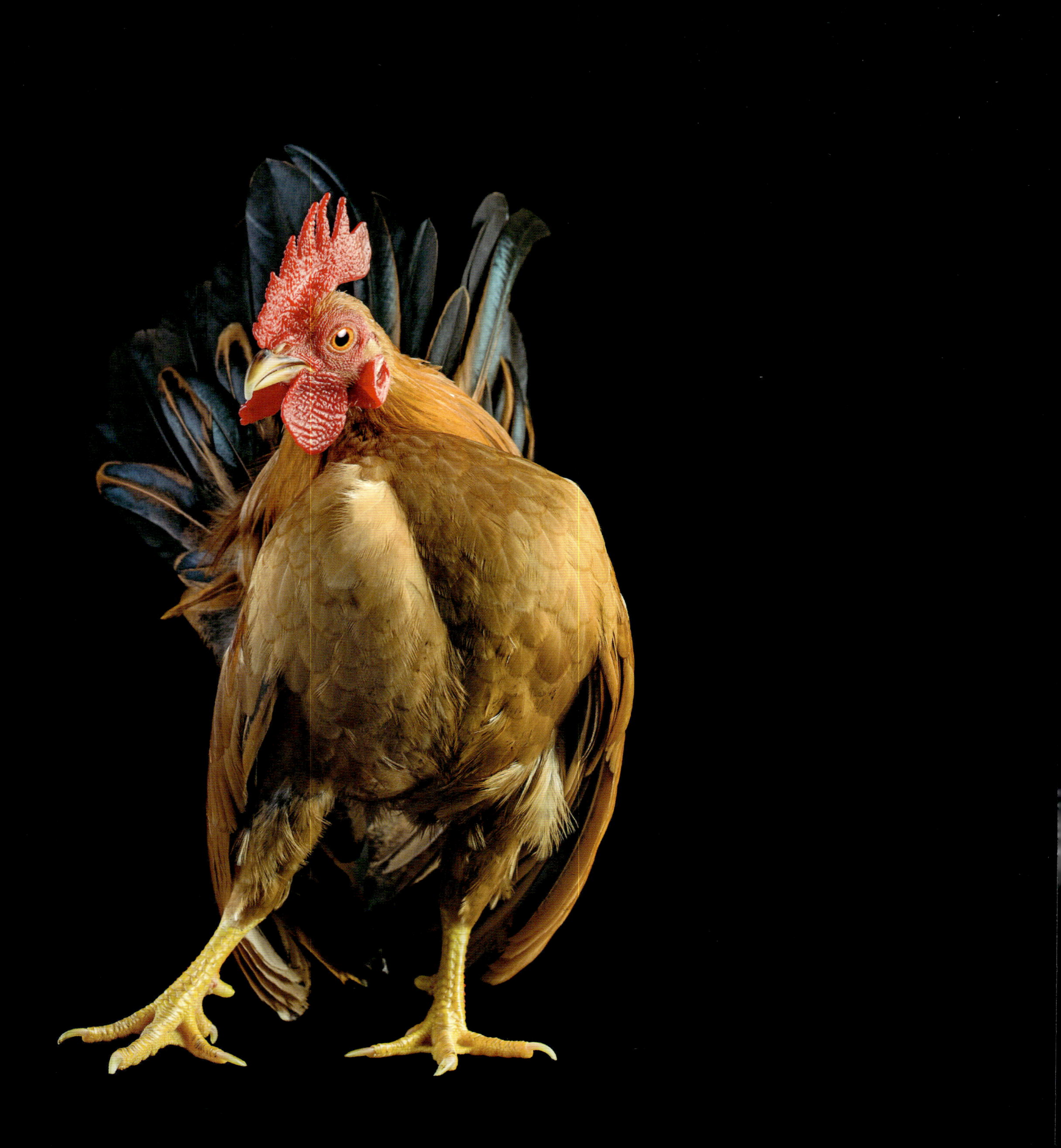

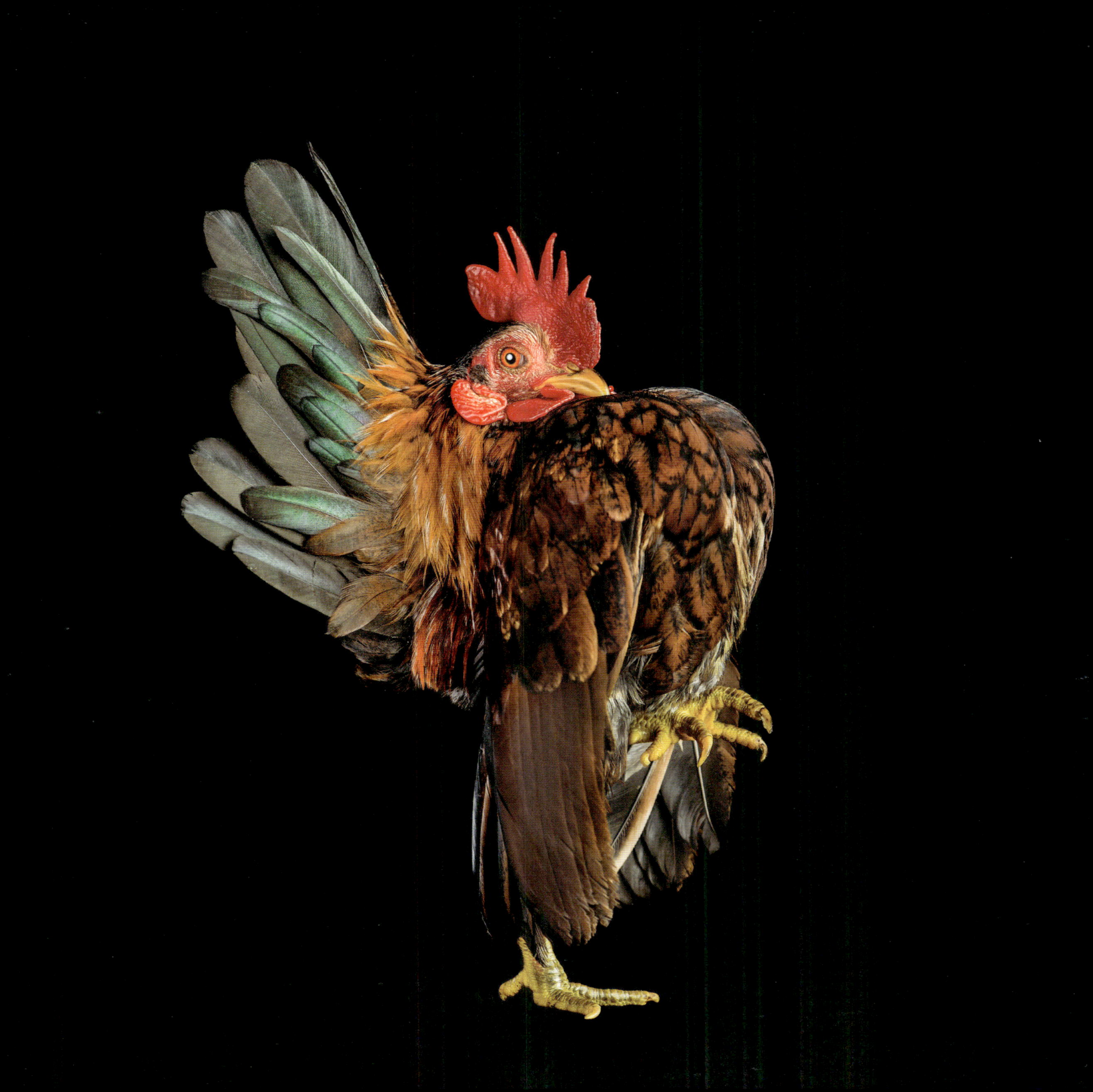

HE WAS
LIKE THE
COCK WHO
THOUGHT
THE SUN HAD
RISEN TO
**HEAR HIM
CROW.**

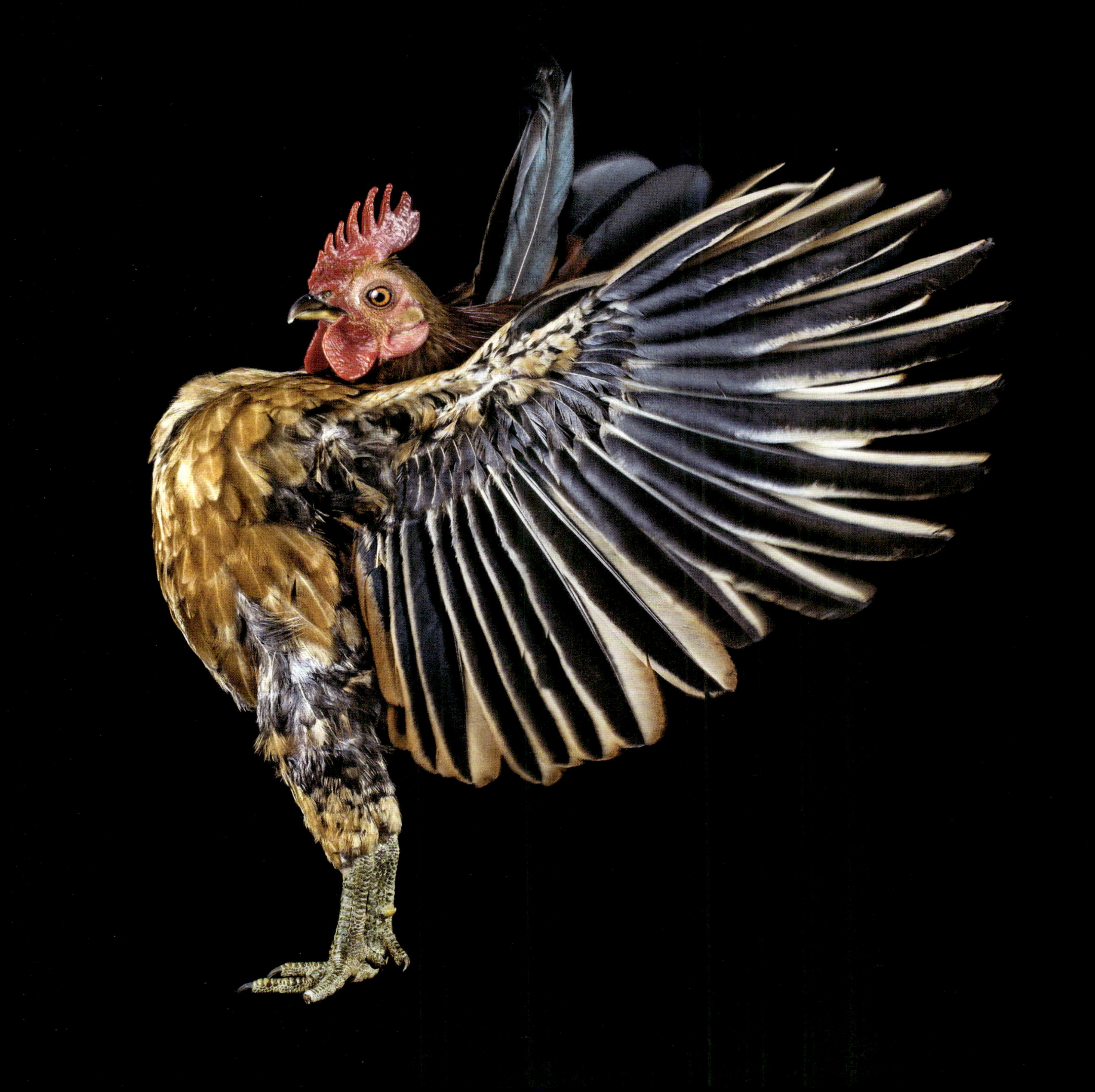

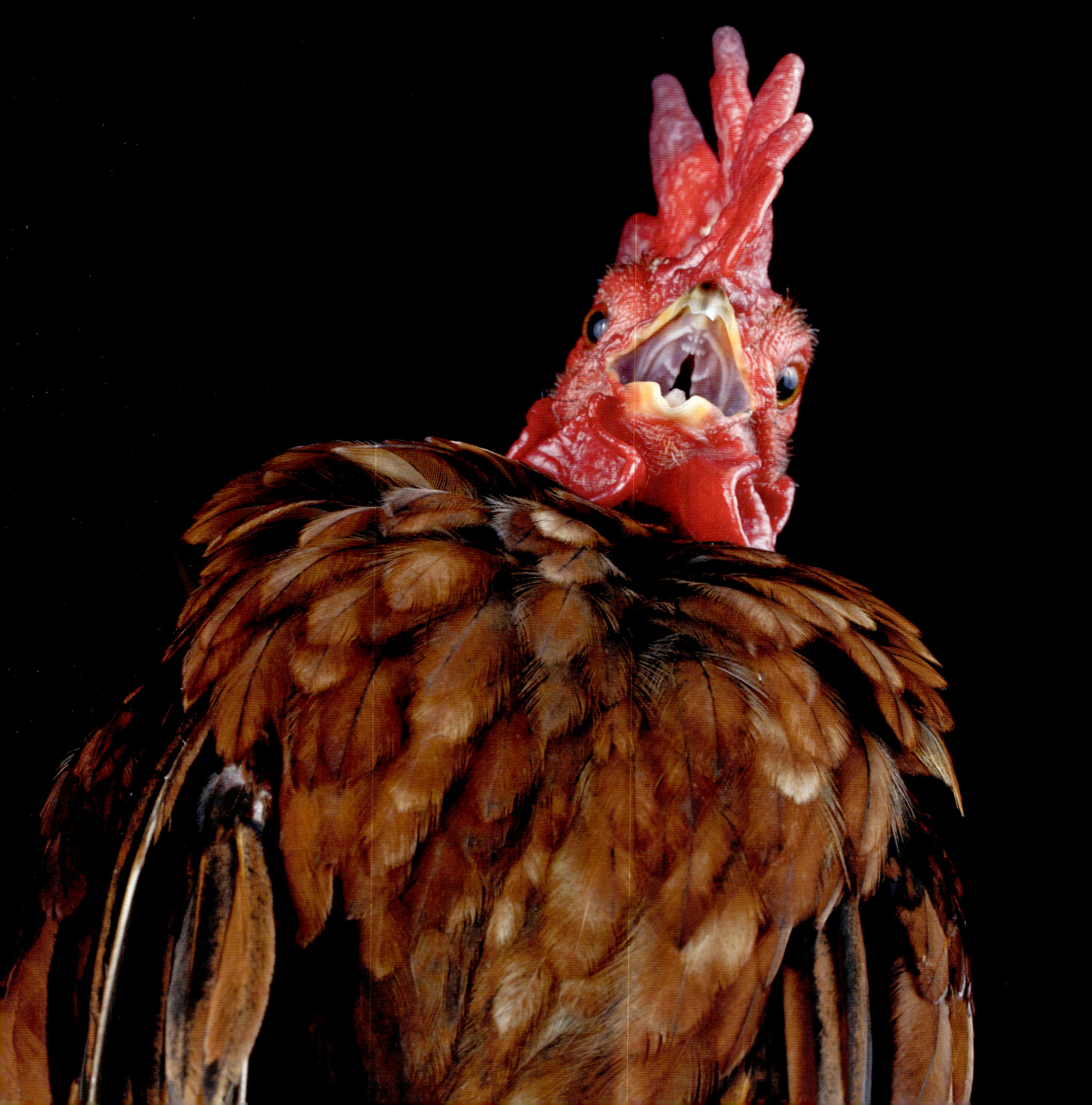

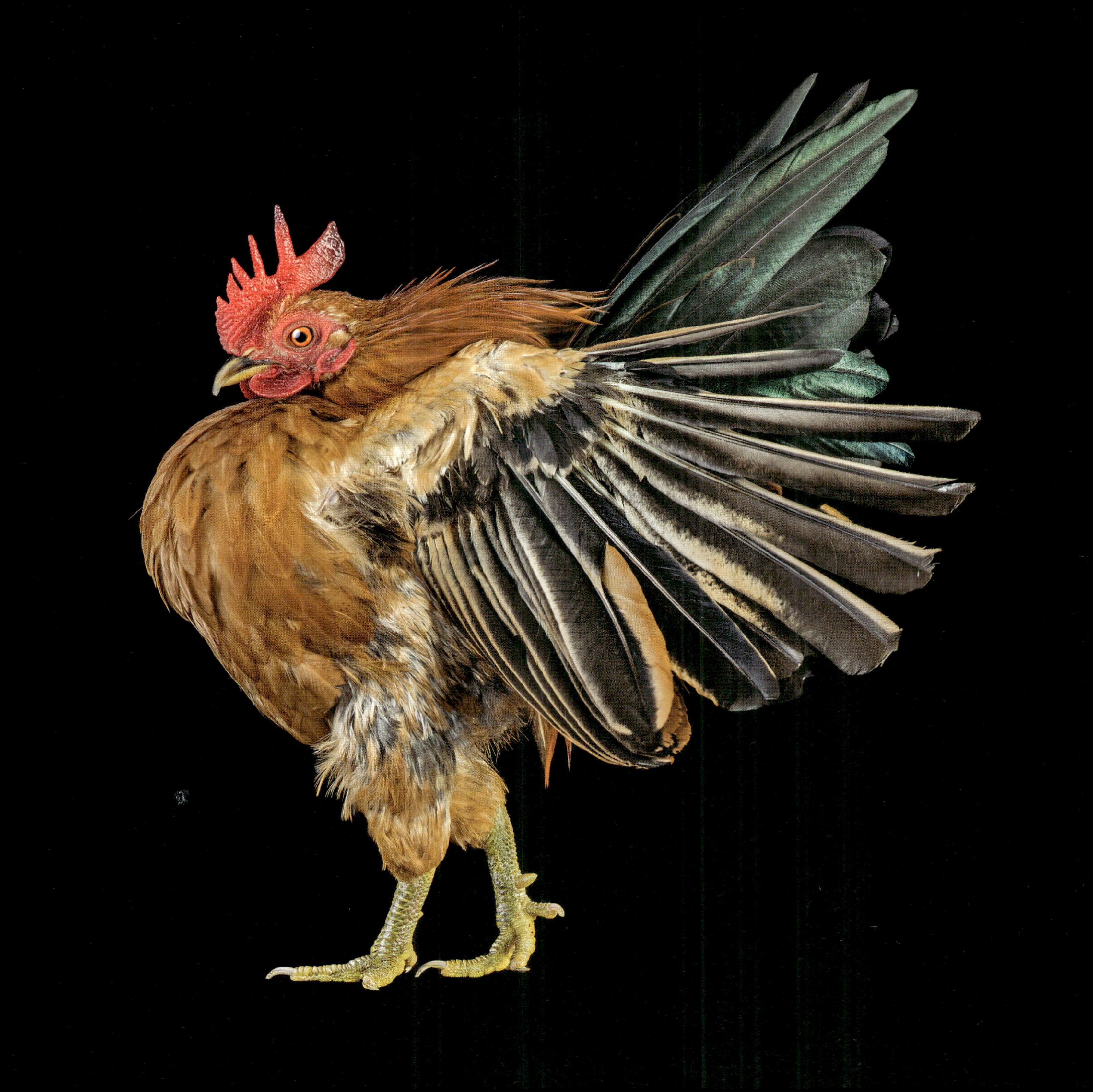

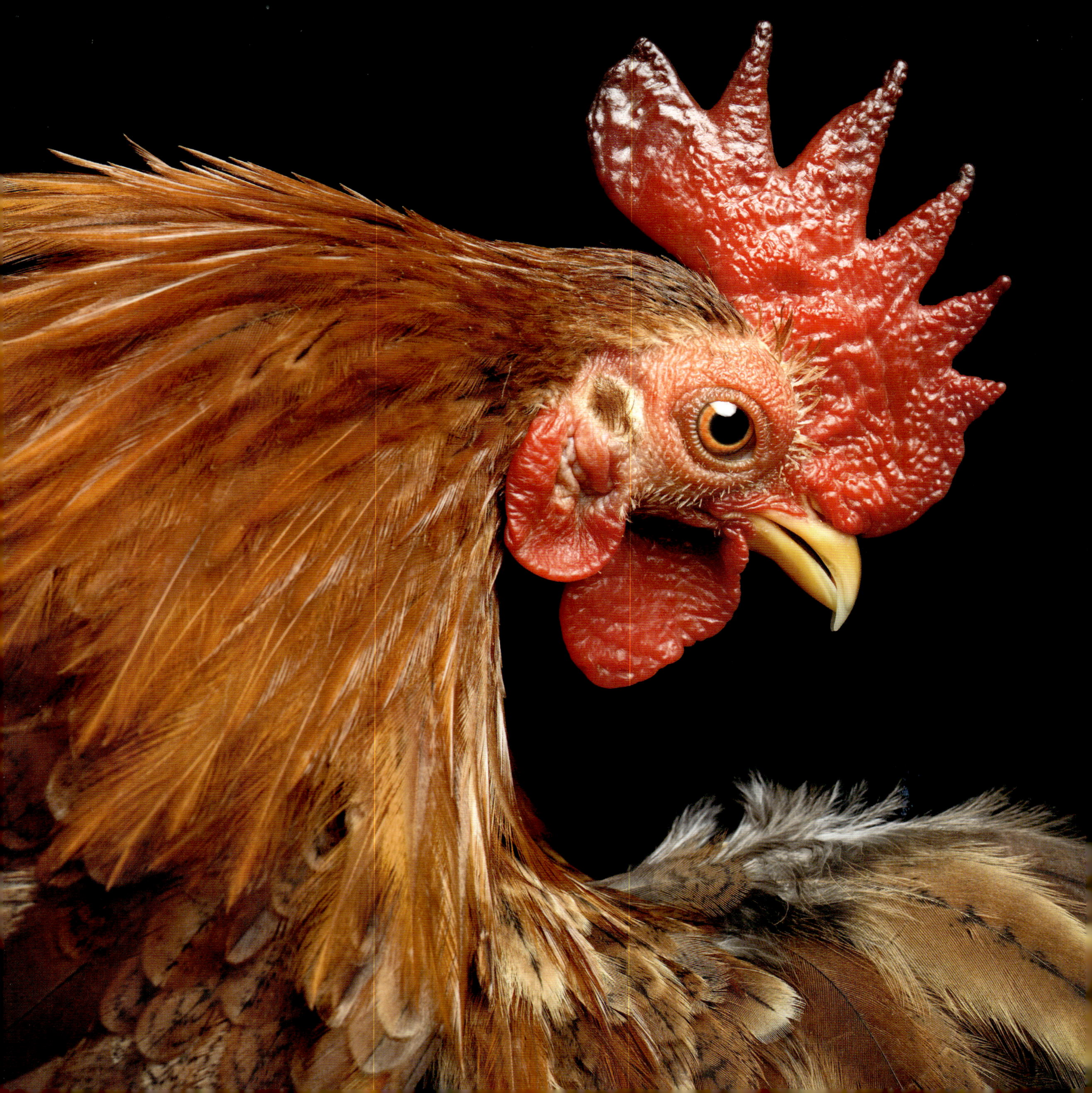

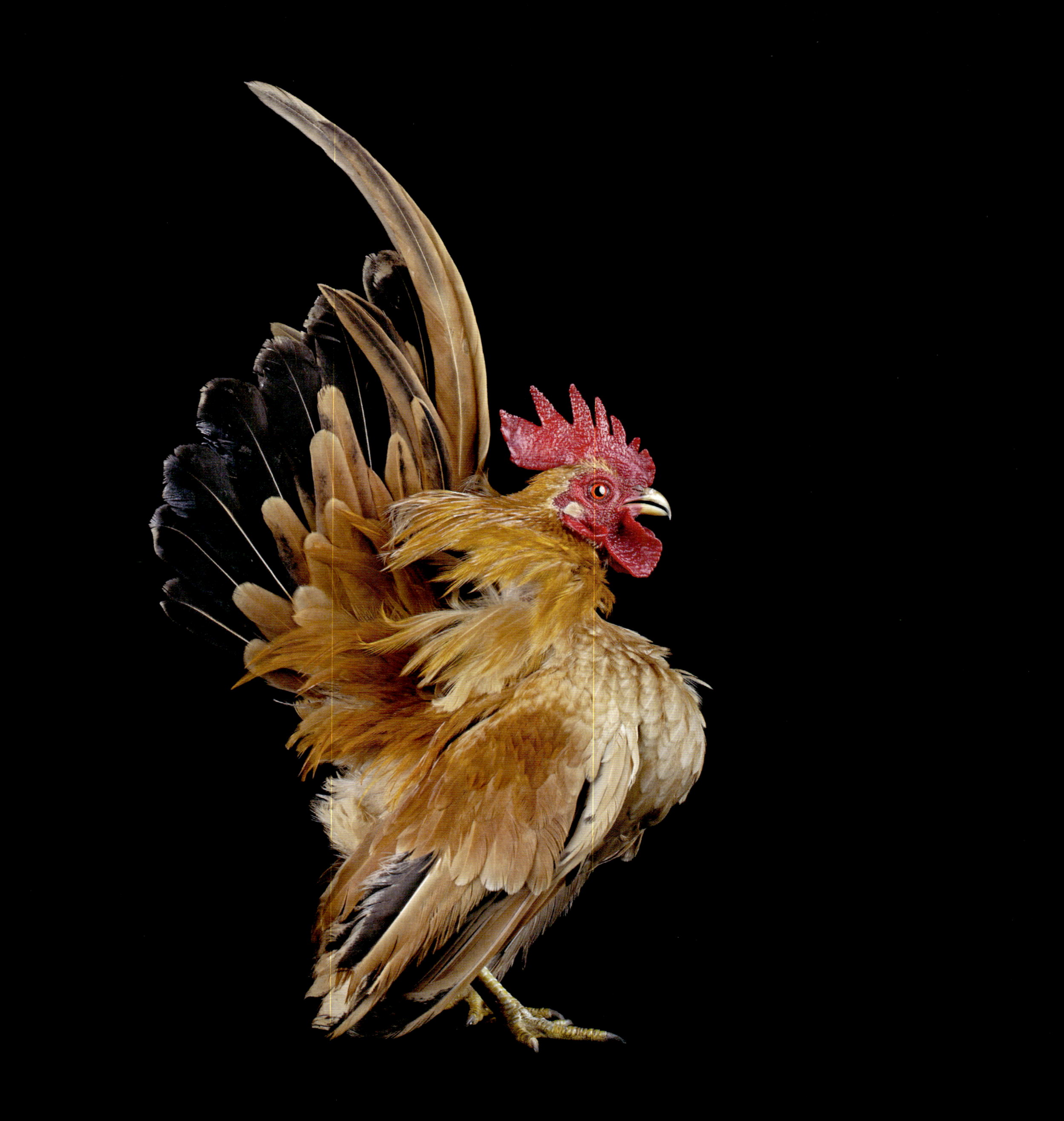

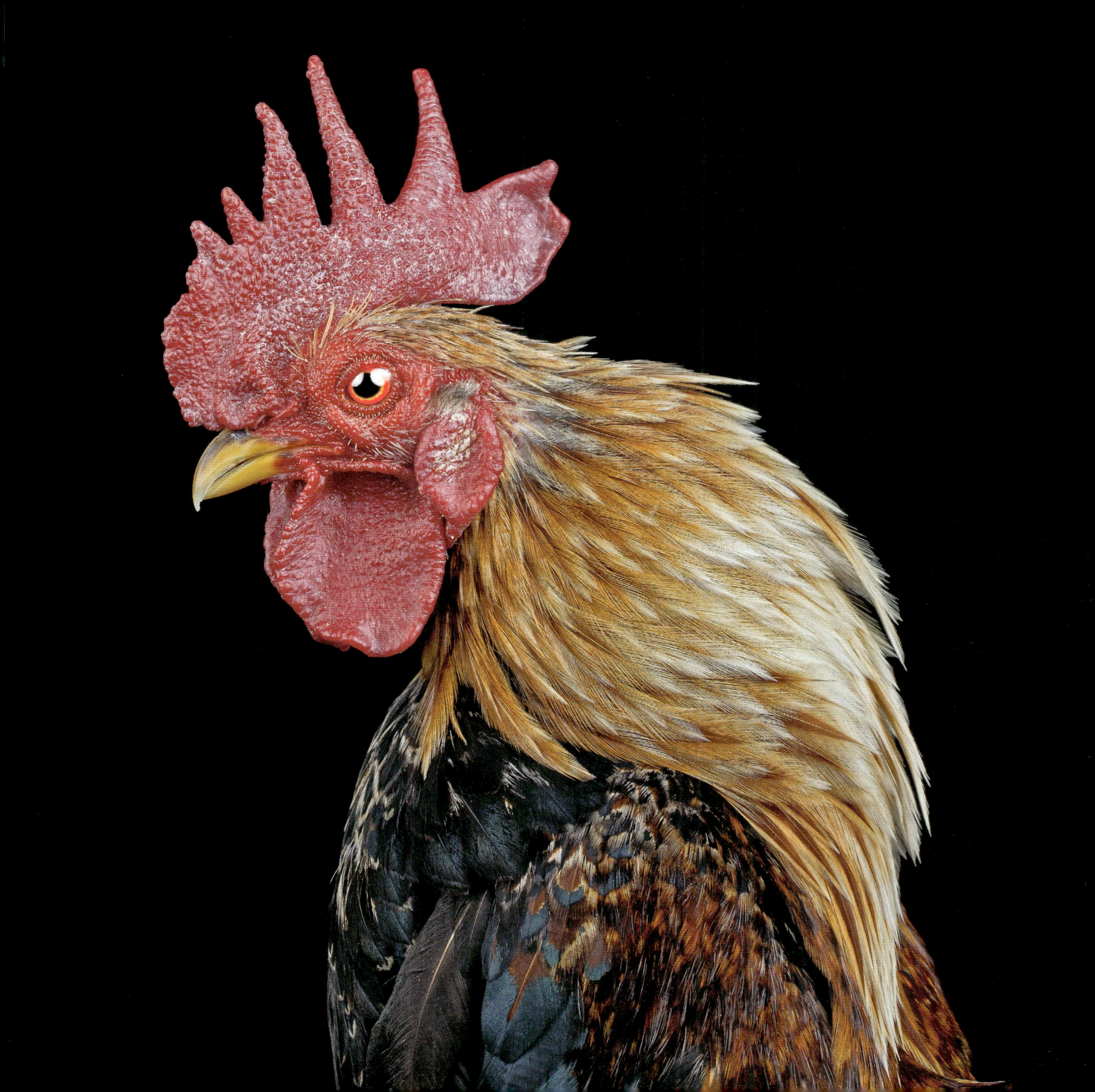

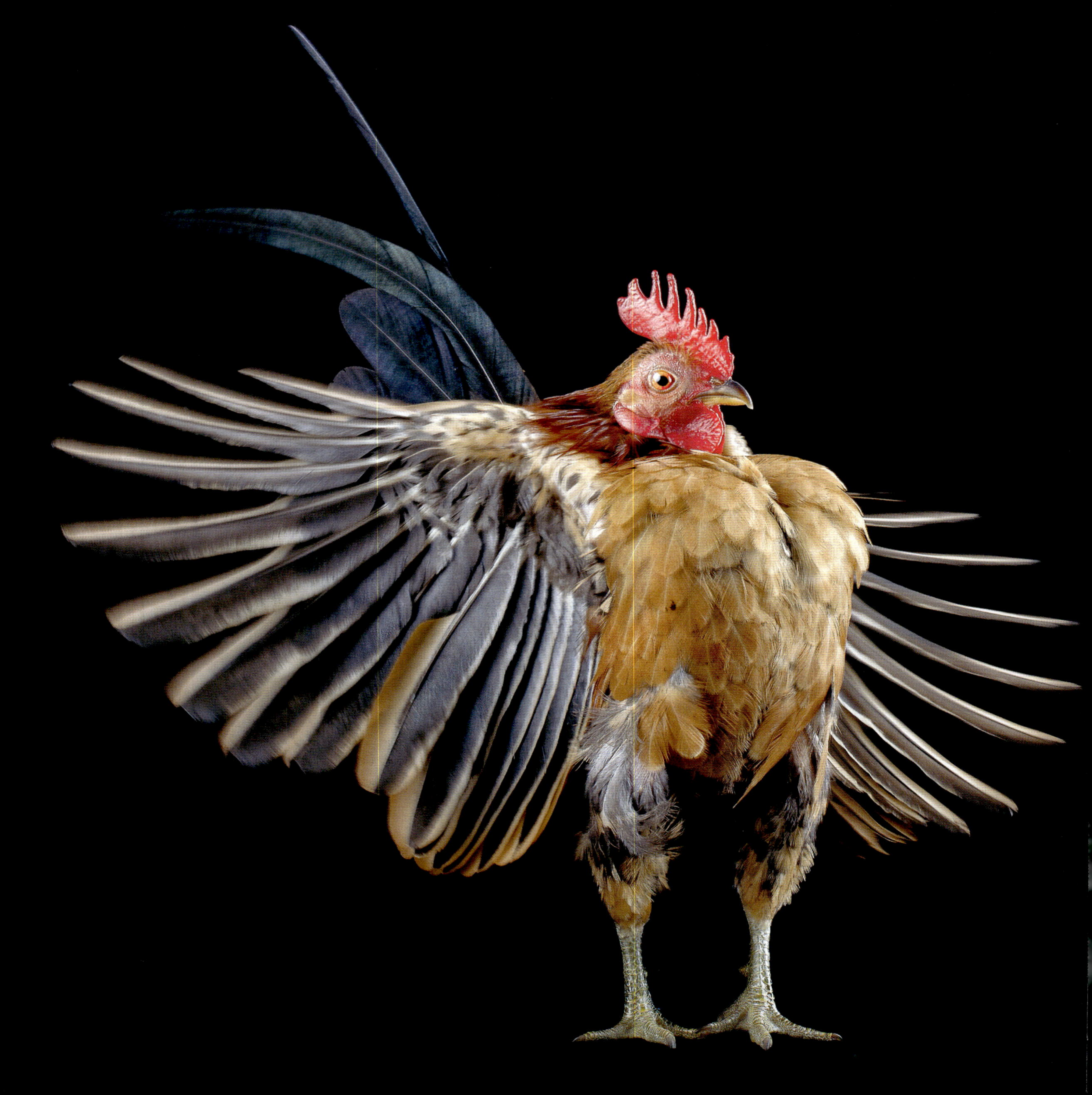

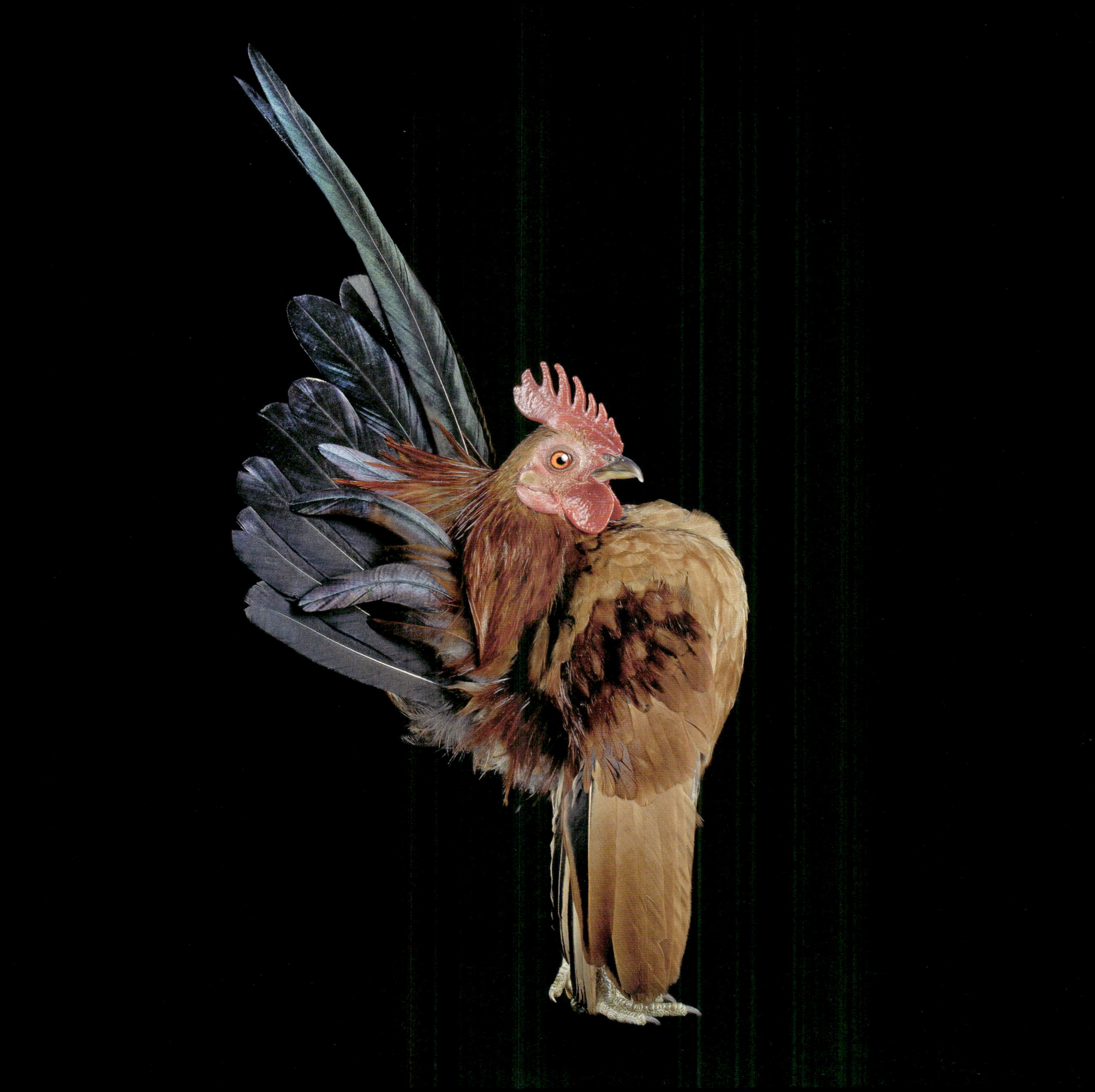

BIRDBRAIN

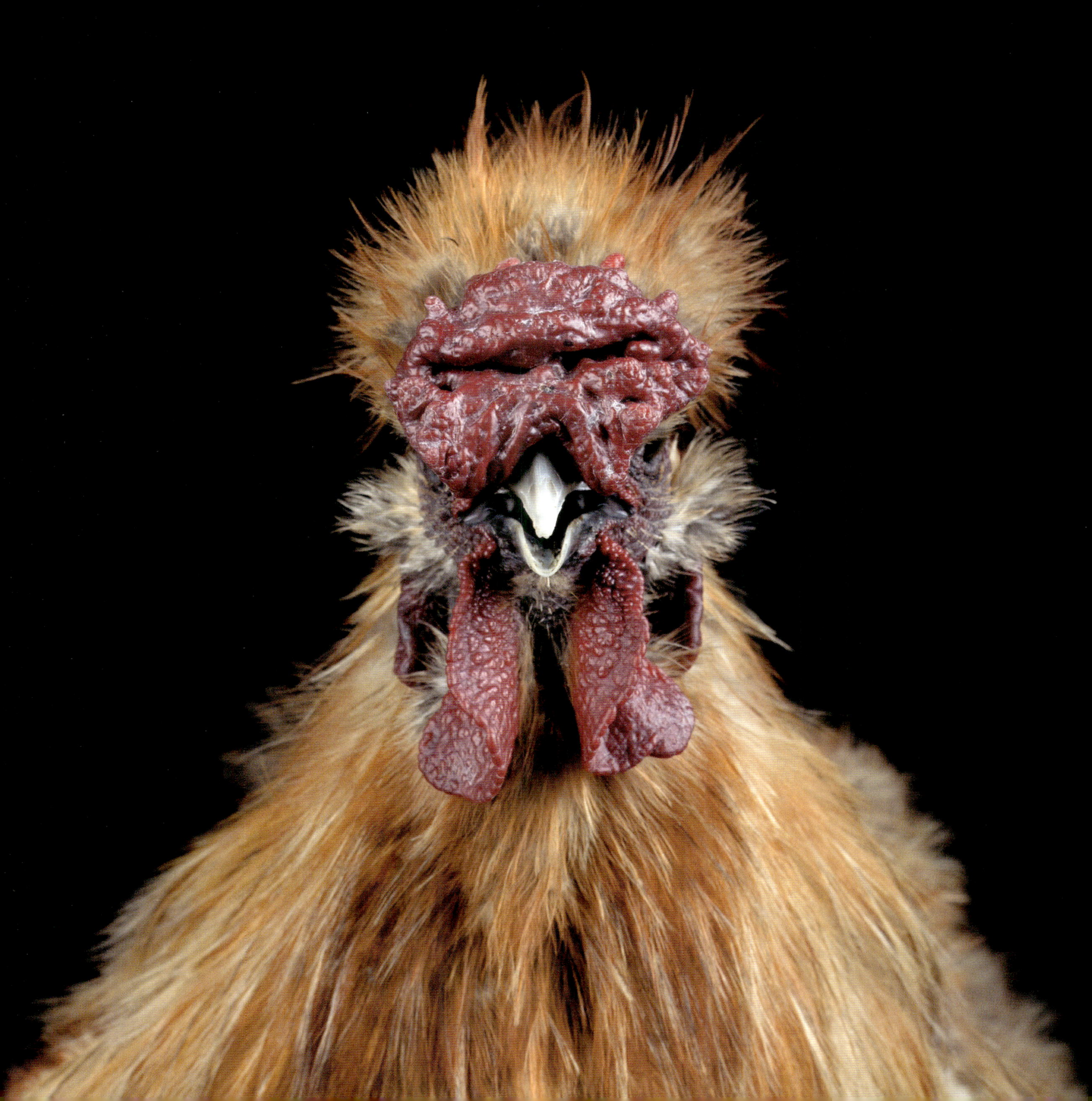

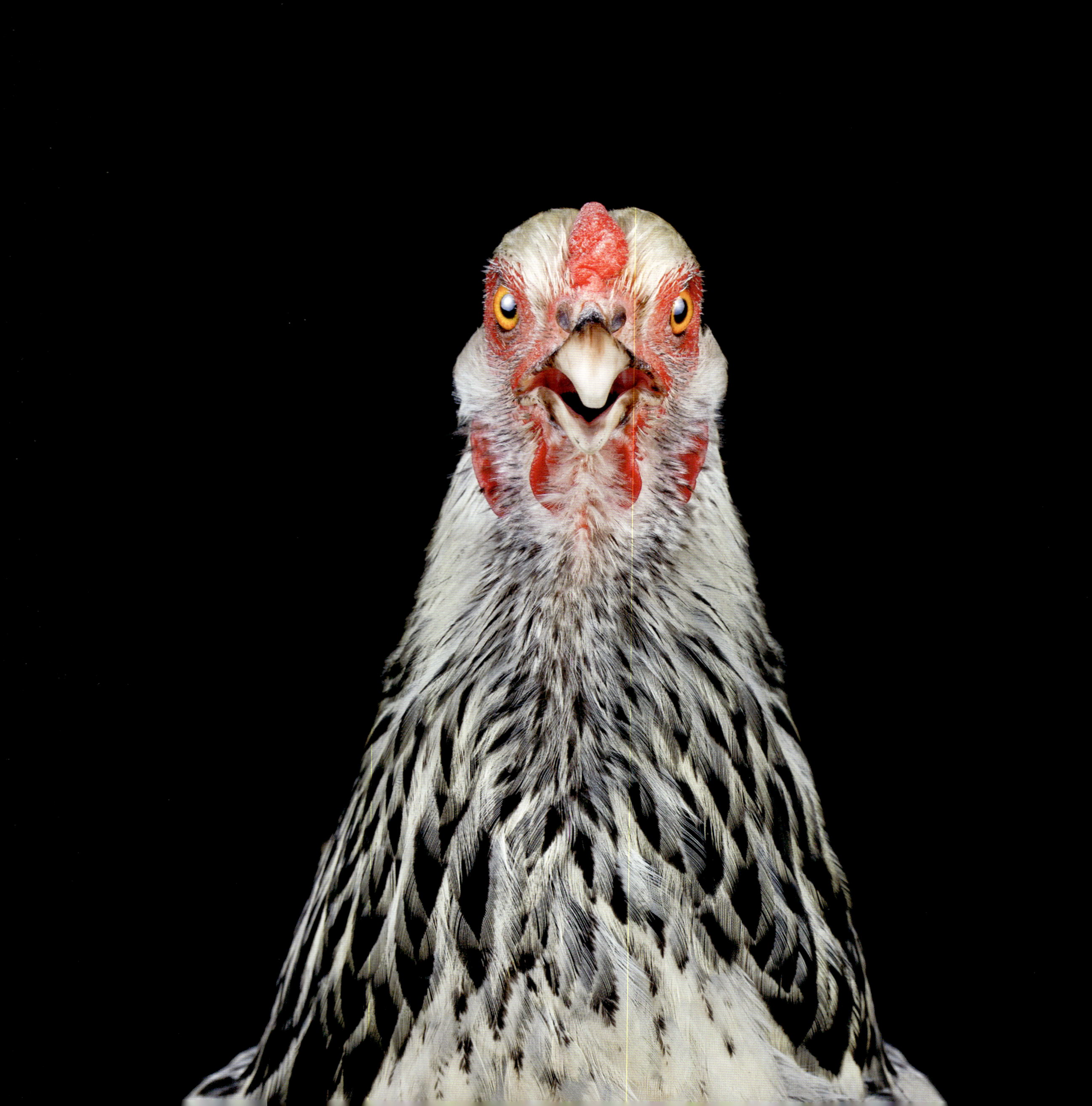

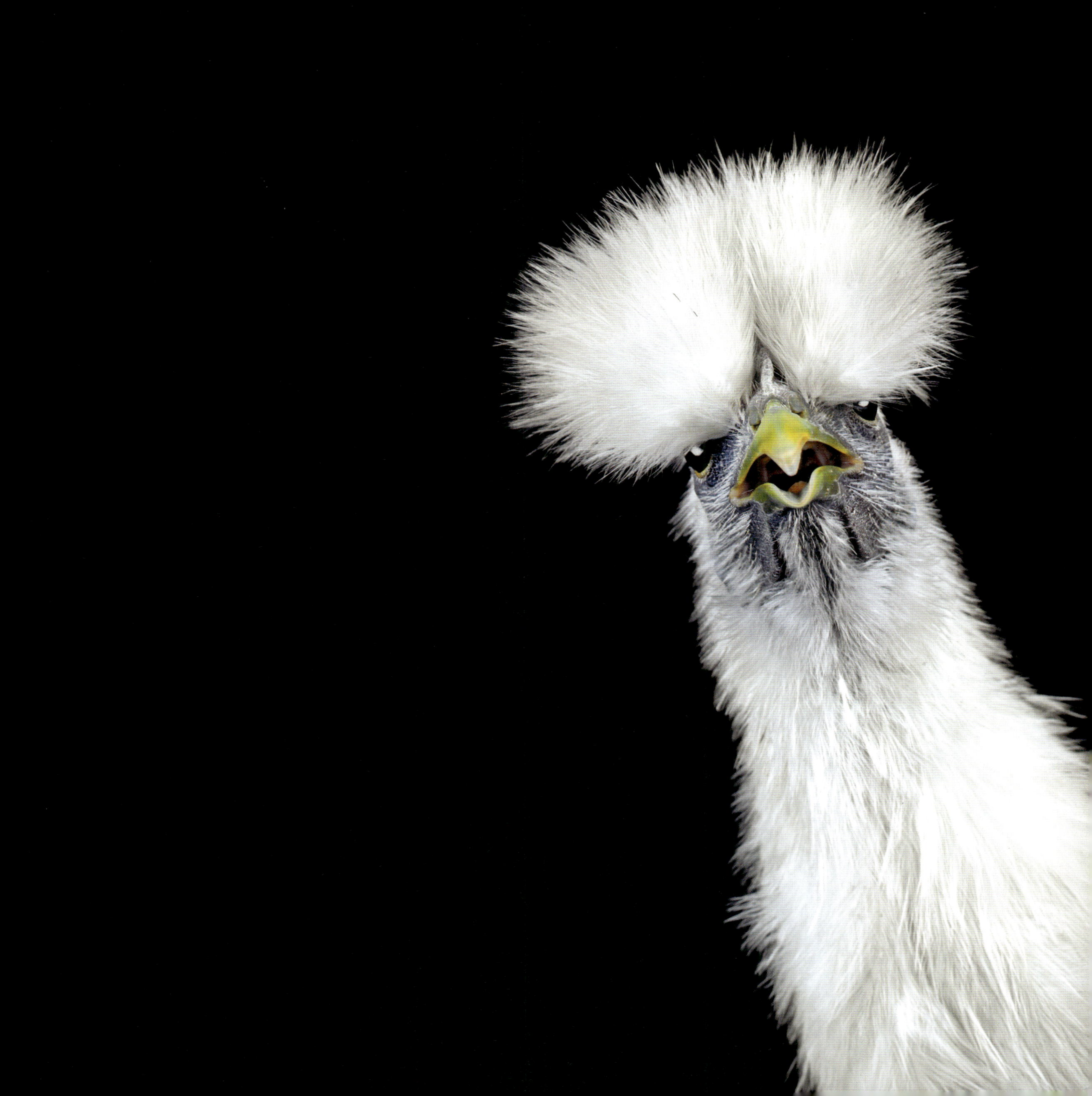

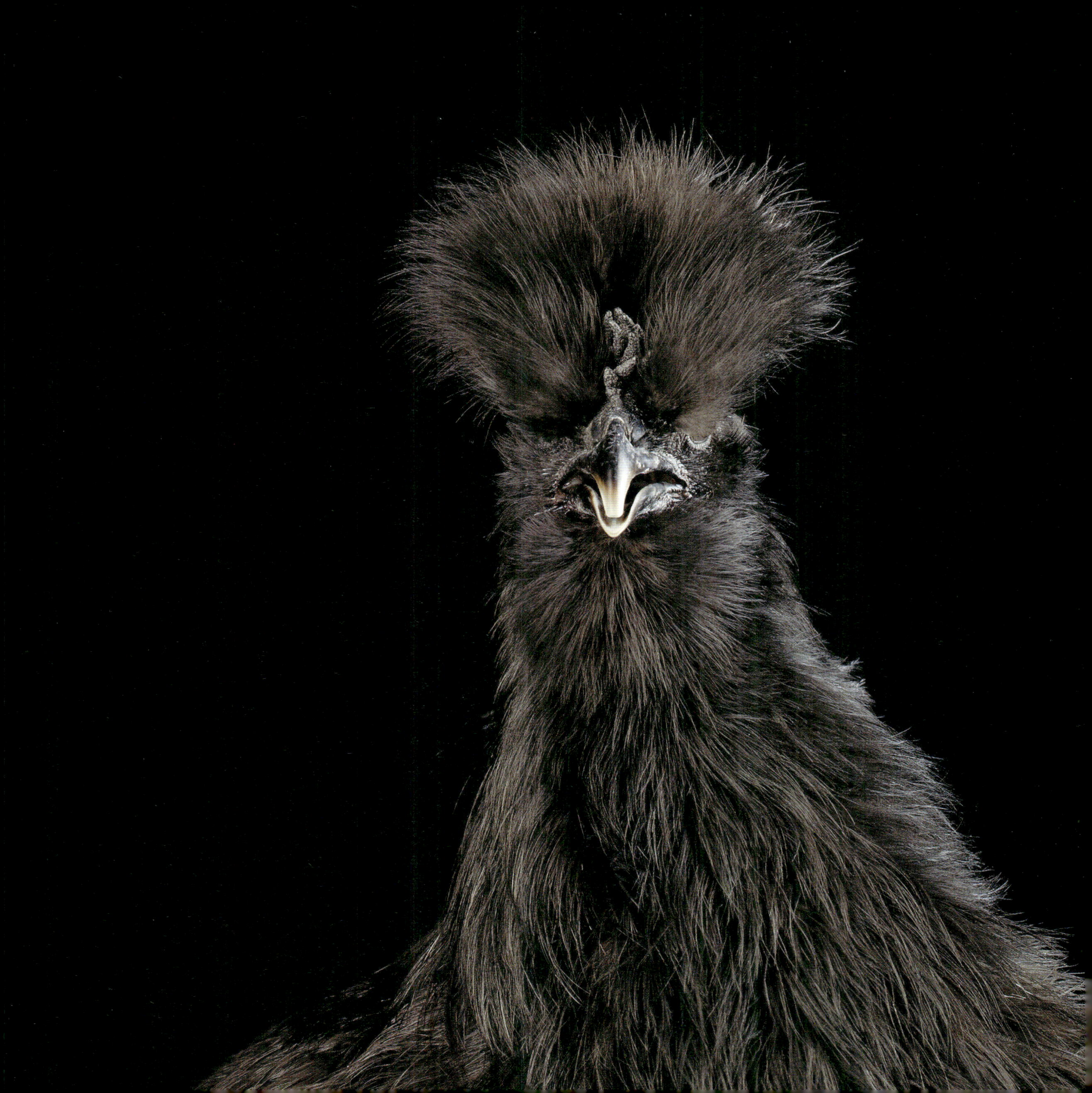

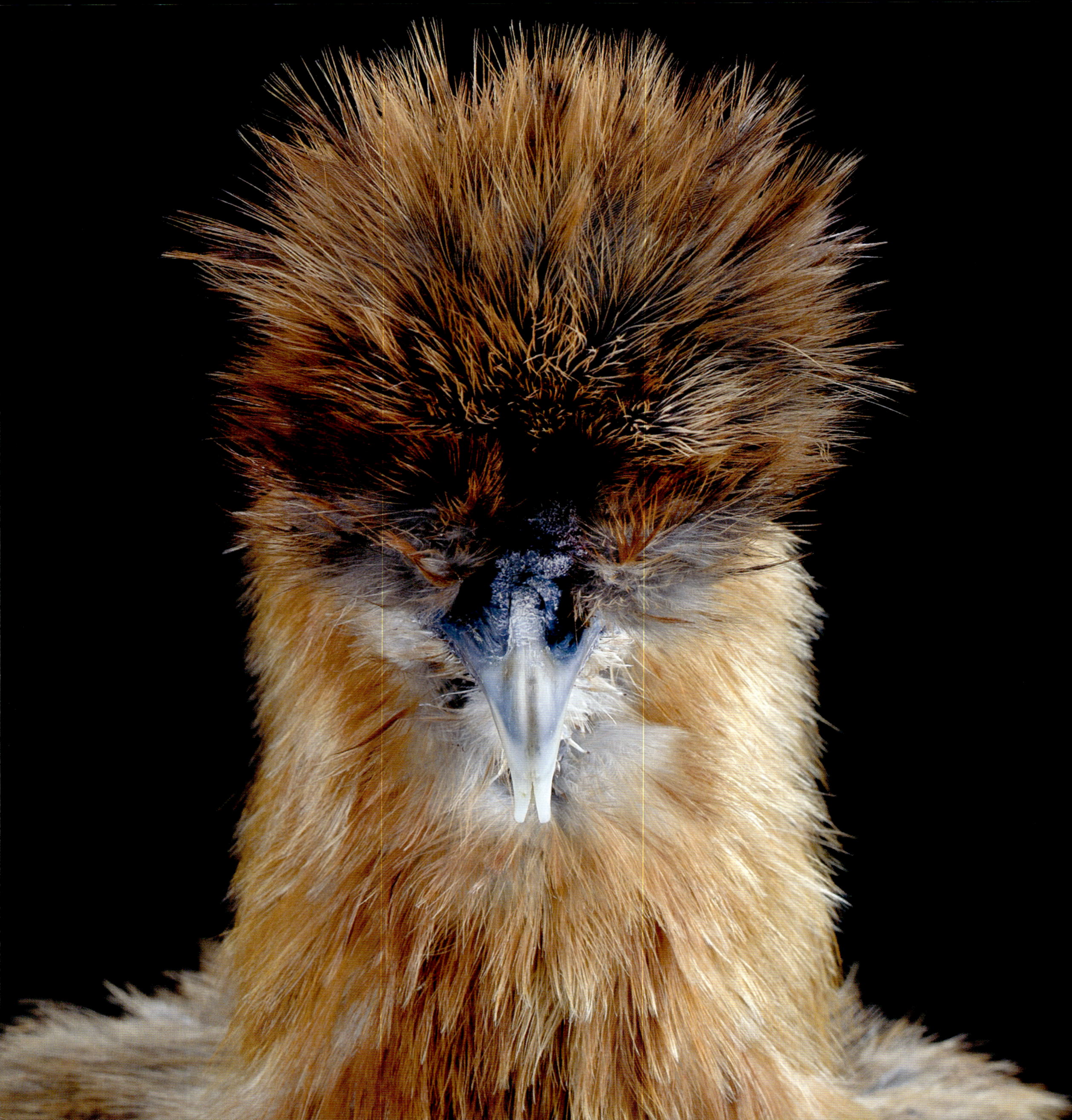

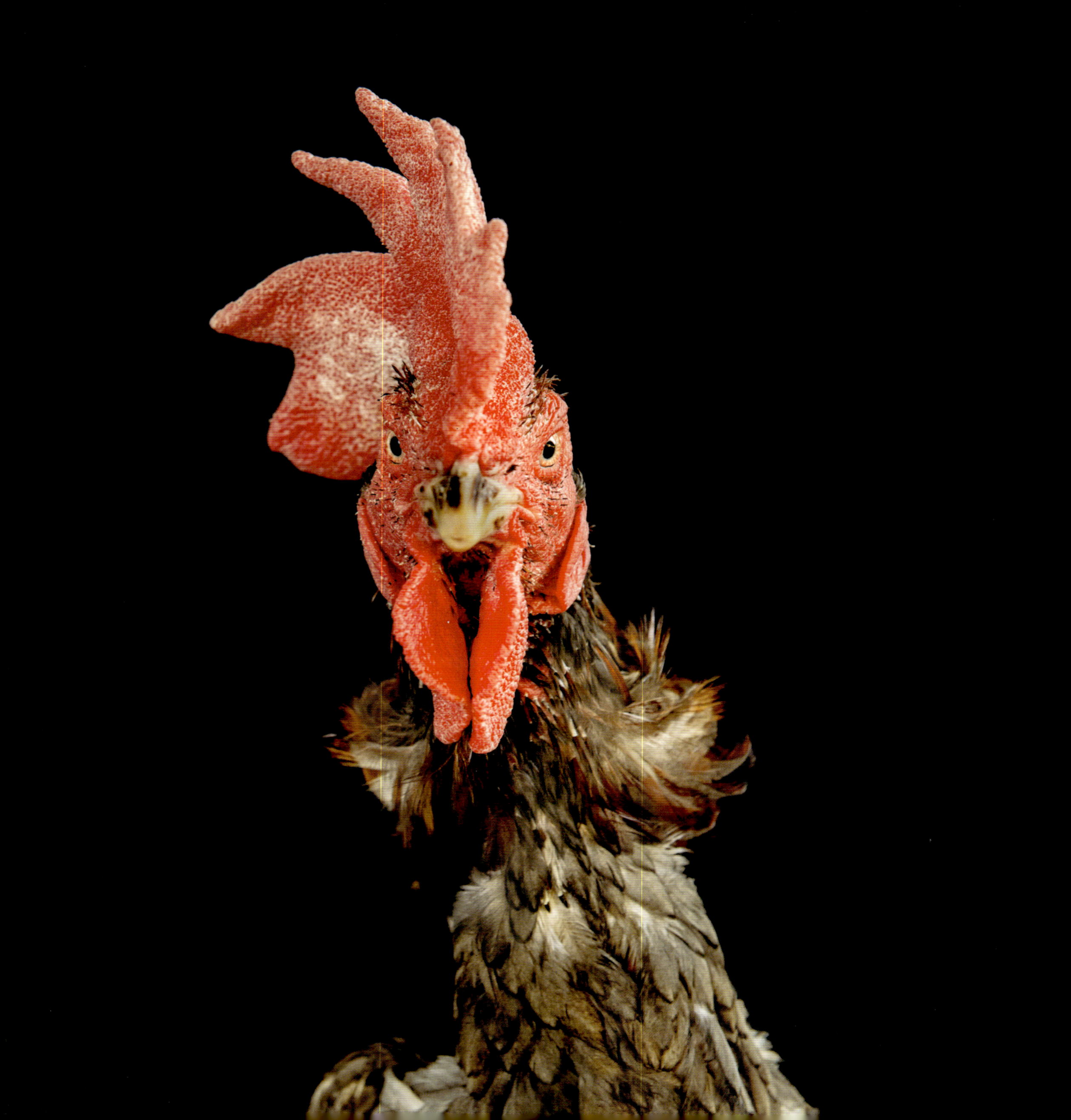

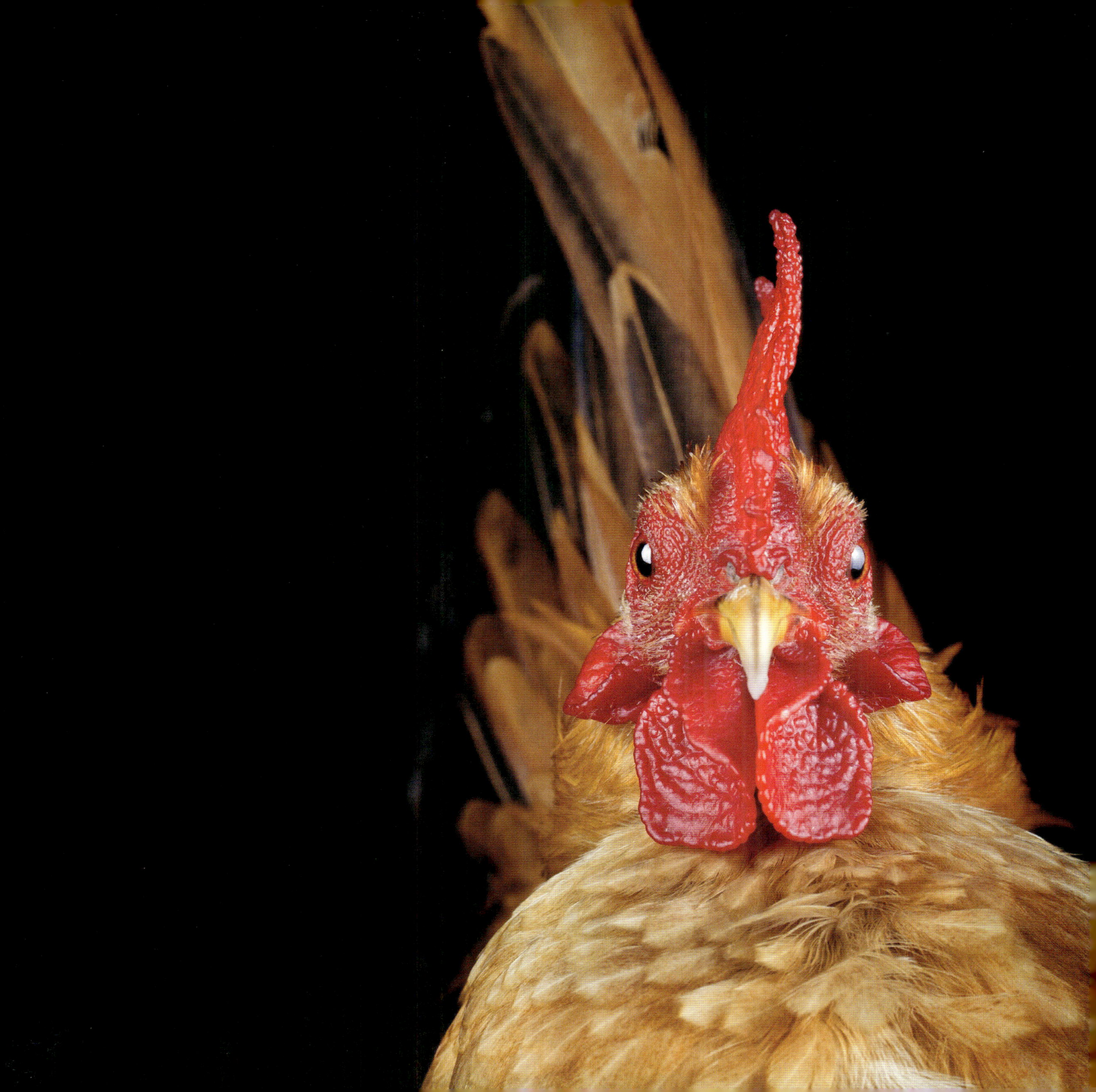

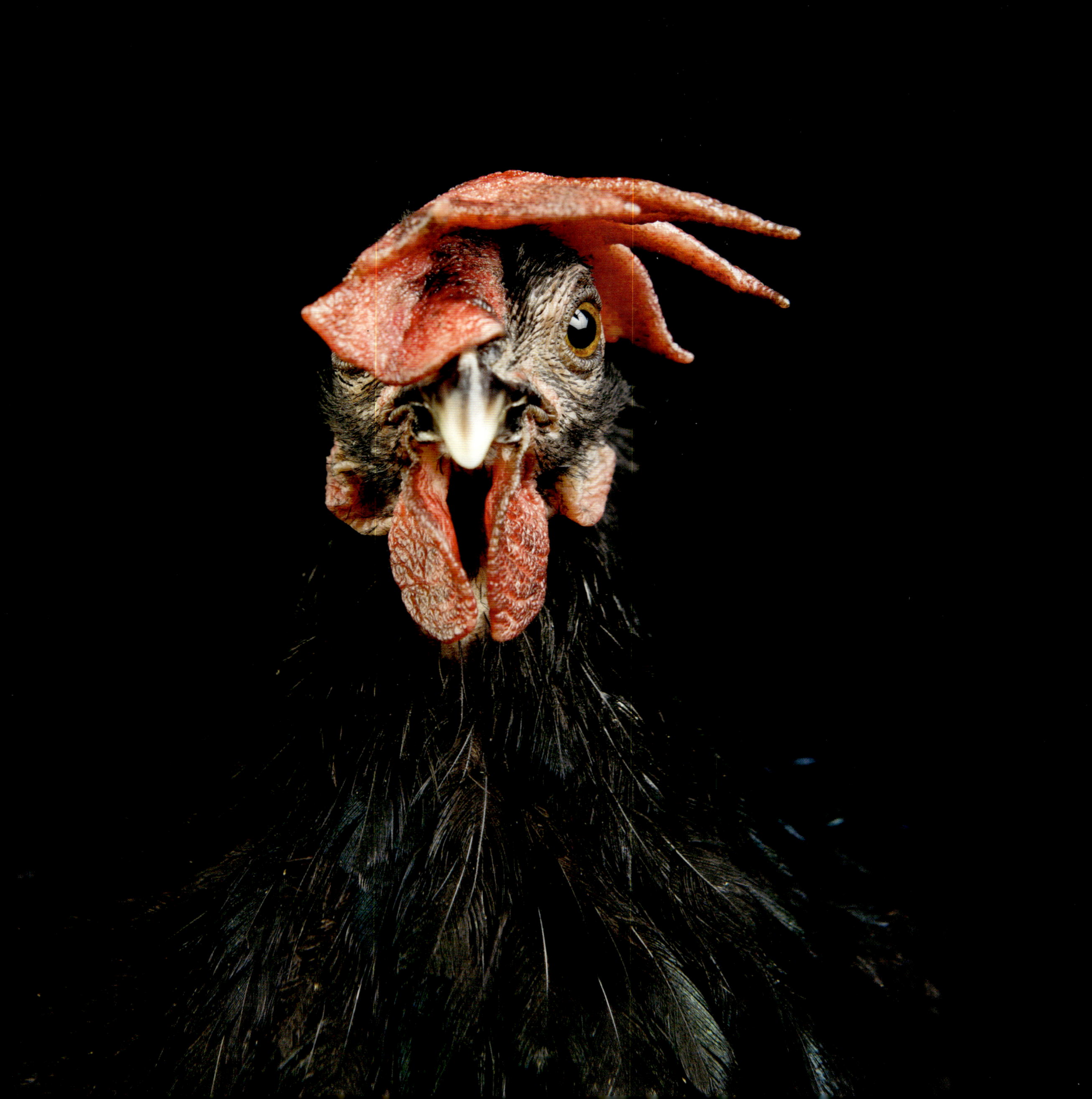

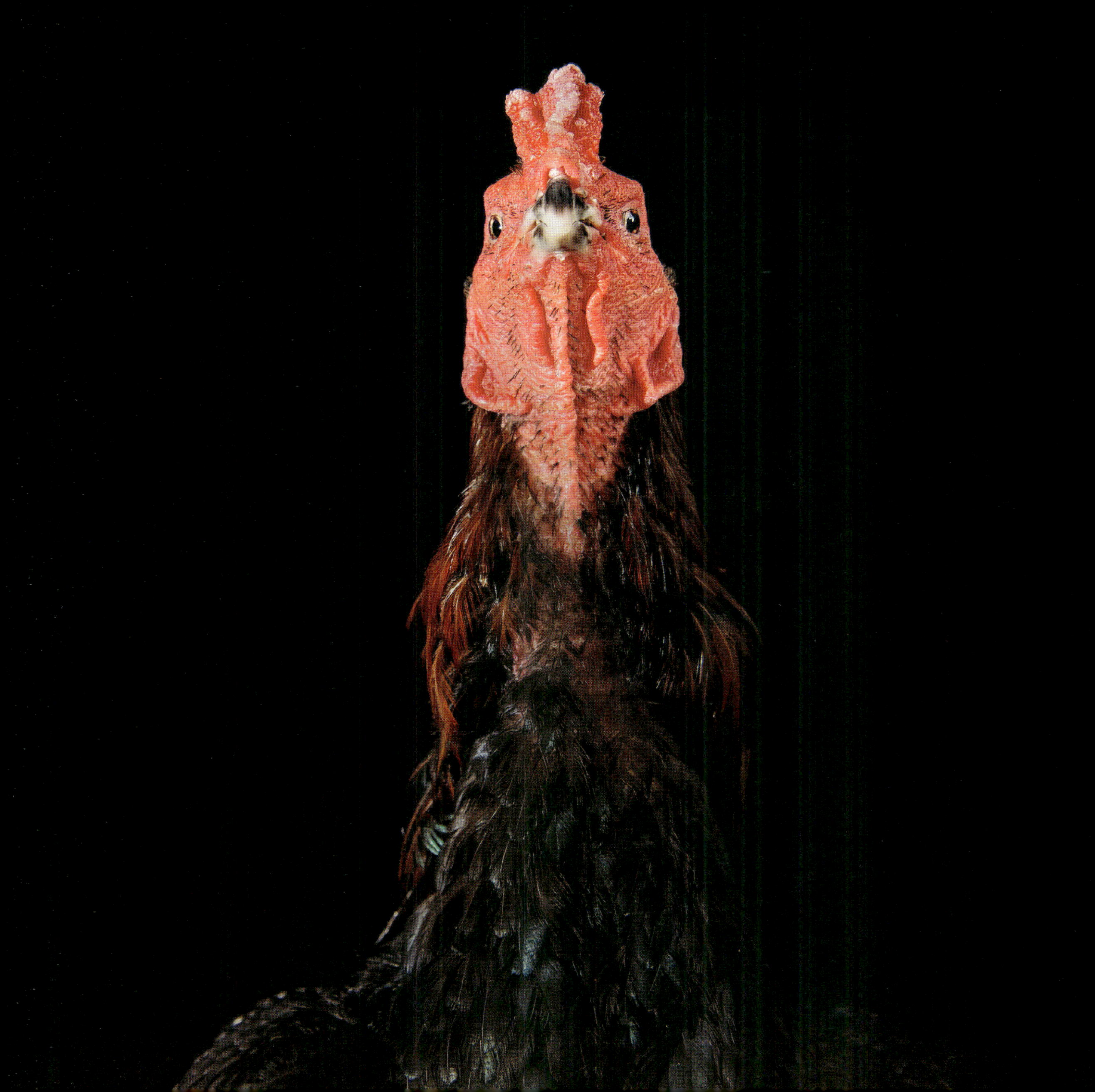

PEOPLE WHO COUNT
THEIR CHICKENS BEFORE
THEY ARE HATCHED
ACT VERY WISELY

BECAUSE CHICKENS RUN ABOUT SO ABSURDLY IT IS IMPOSSIBLE TO COUNT THEM ACCURATELY.

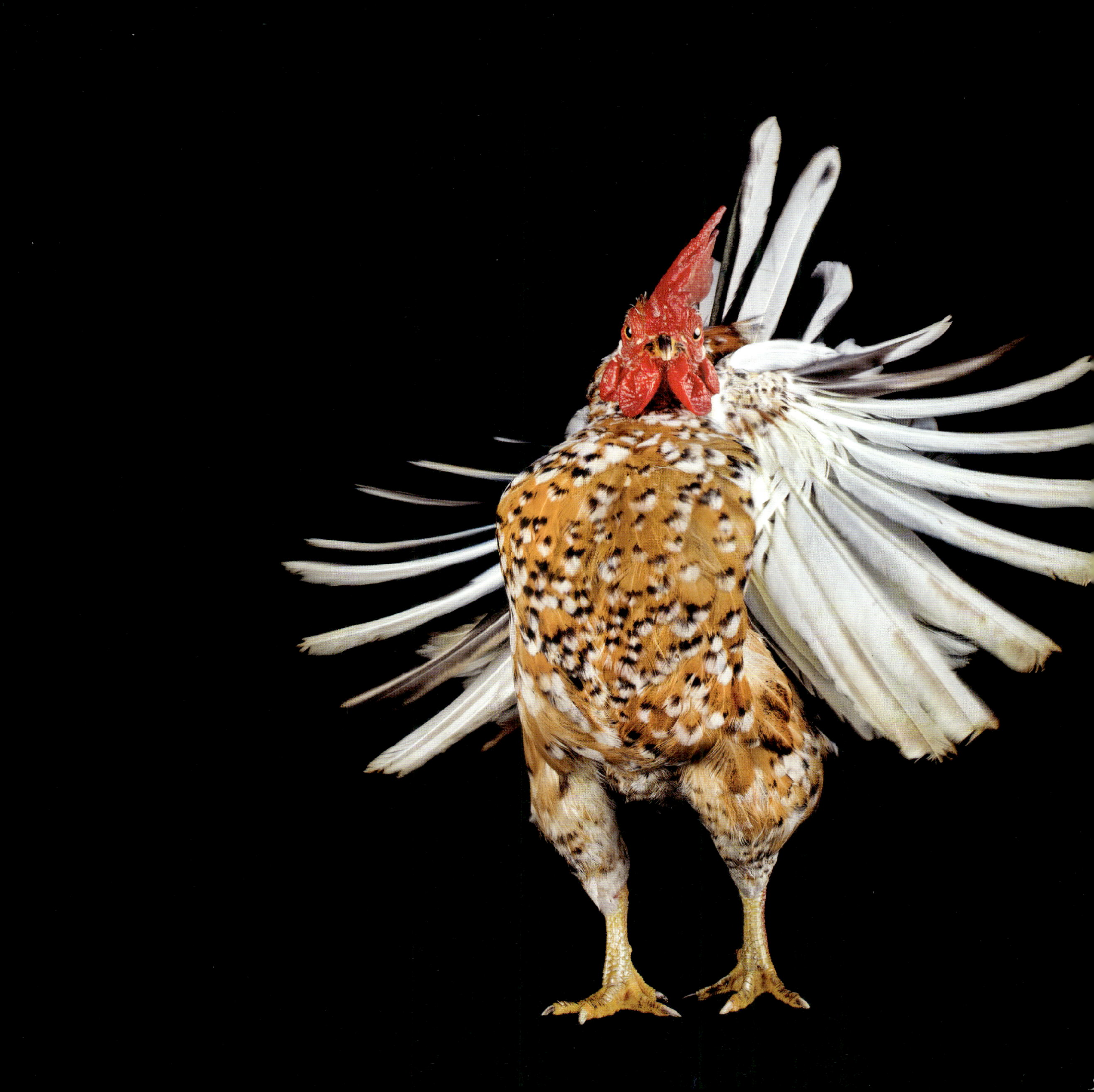

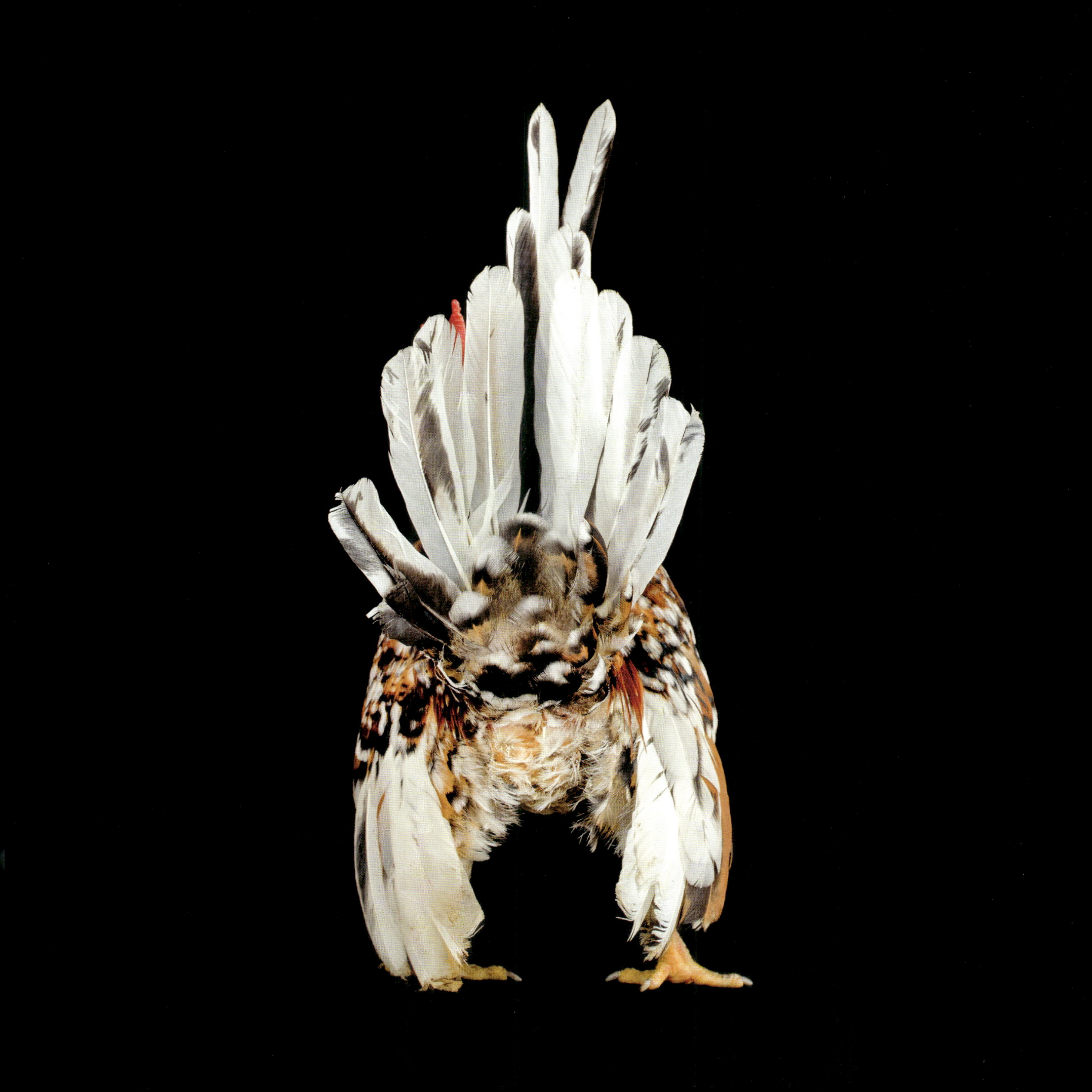

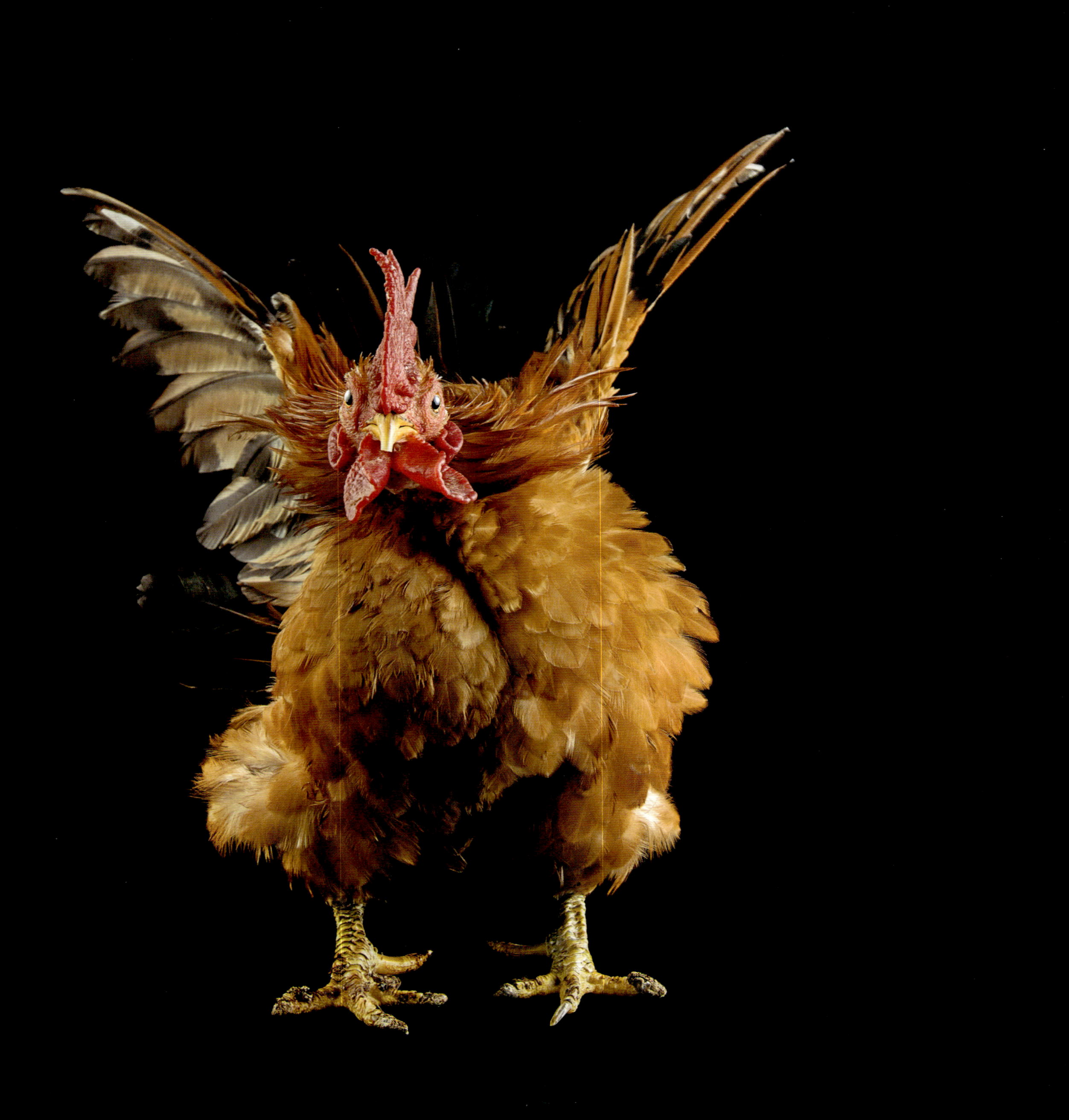

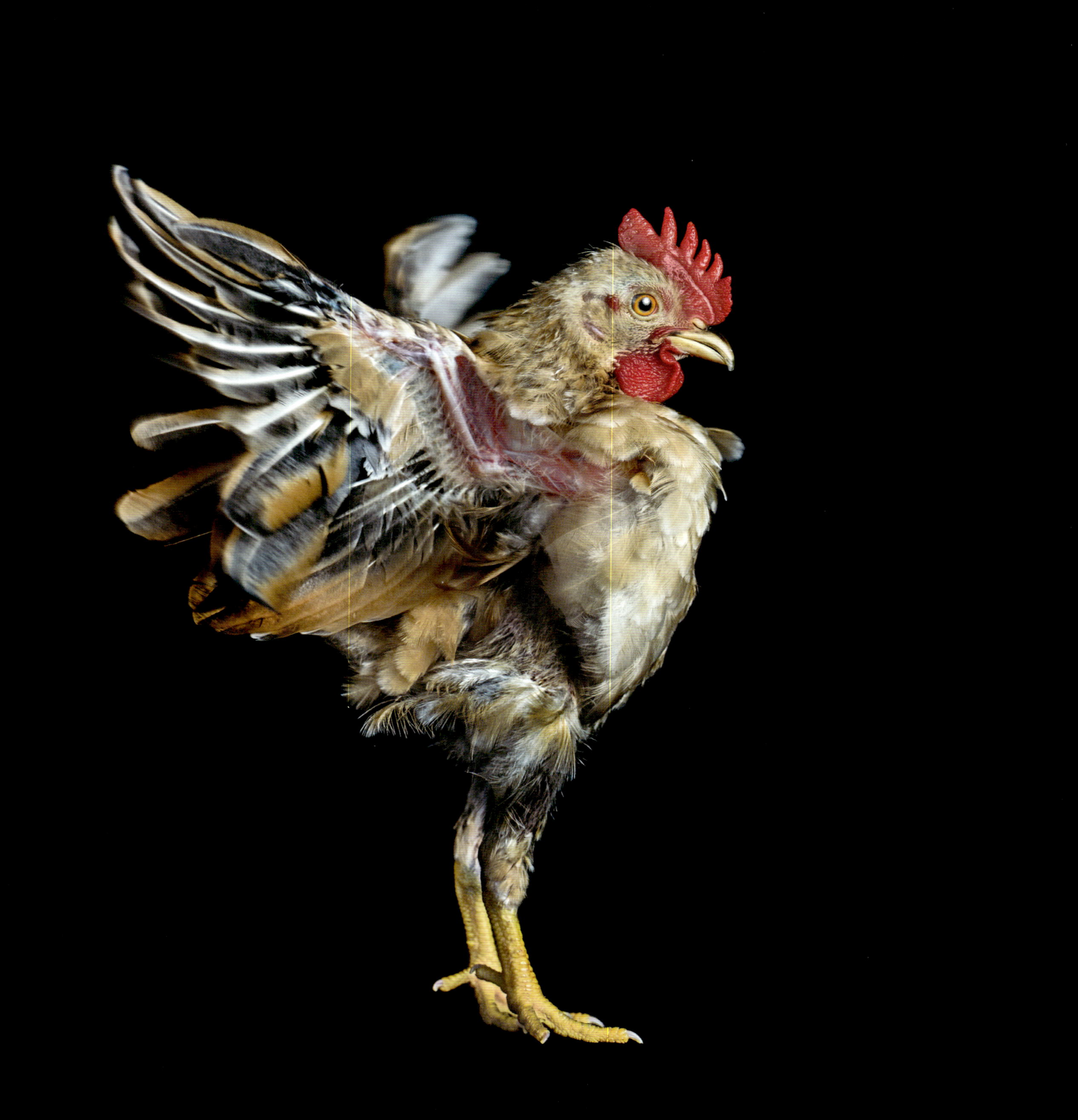

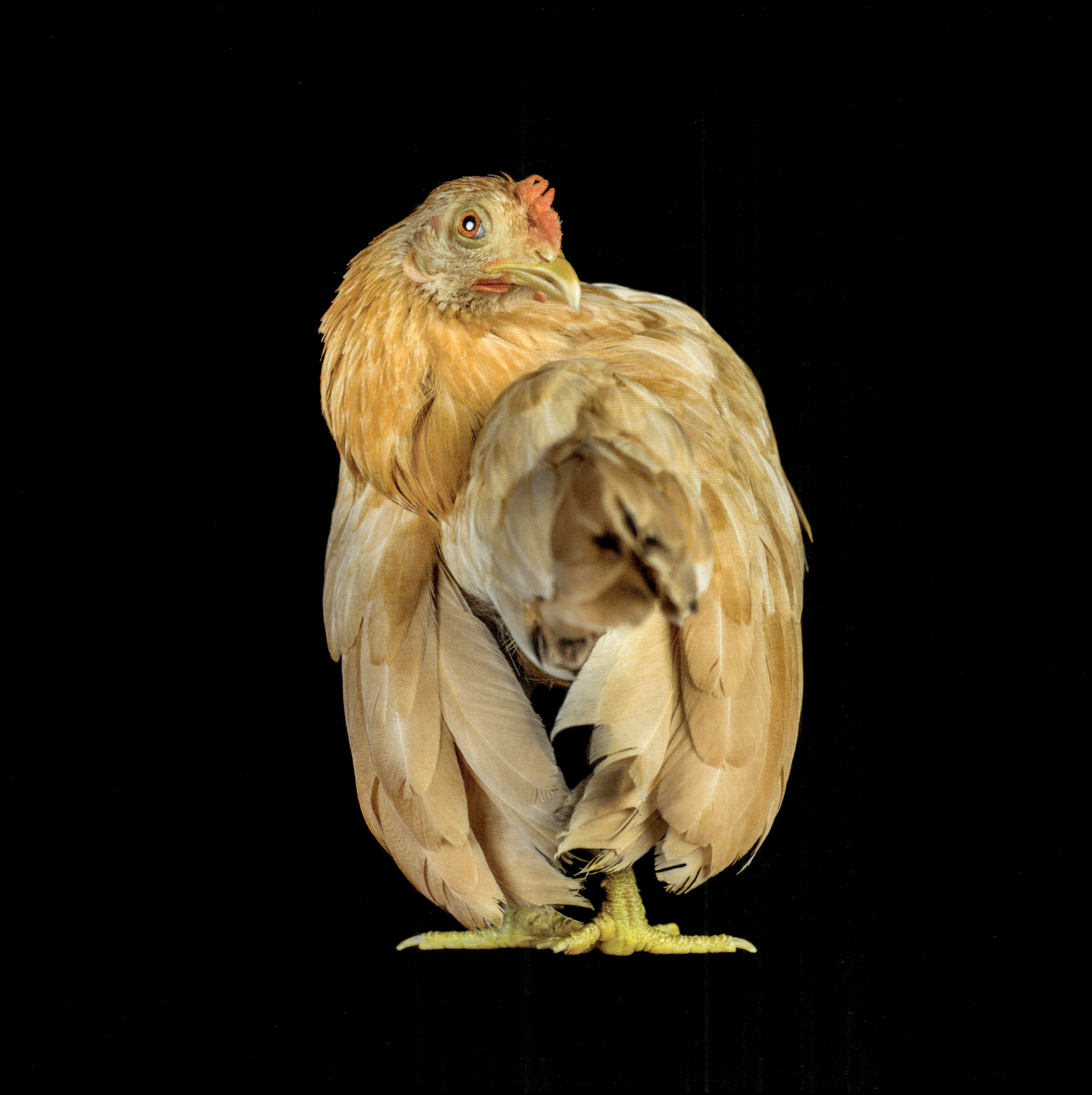

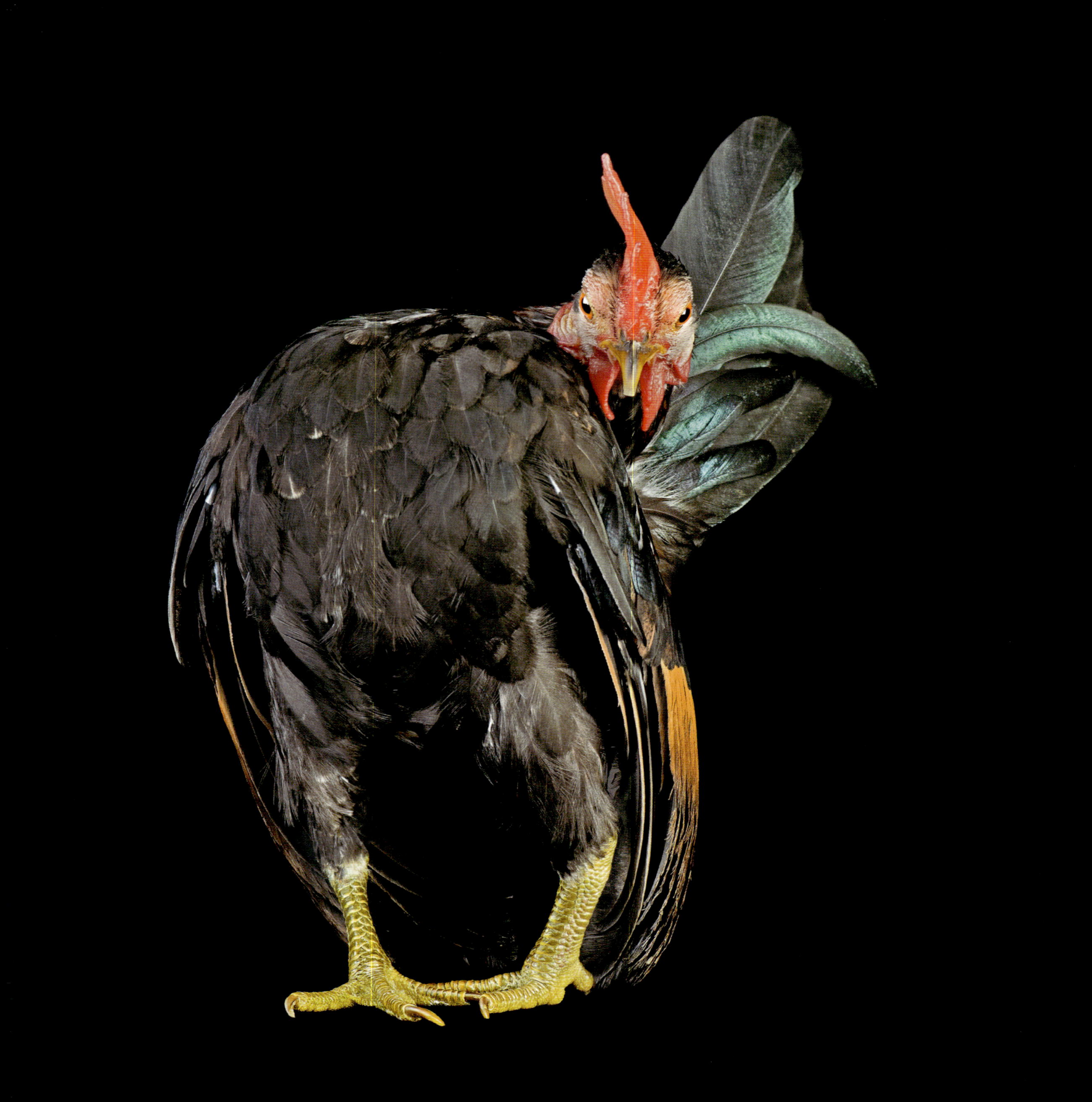

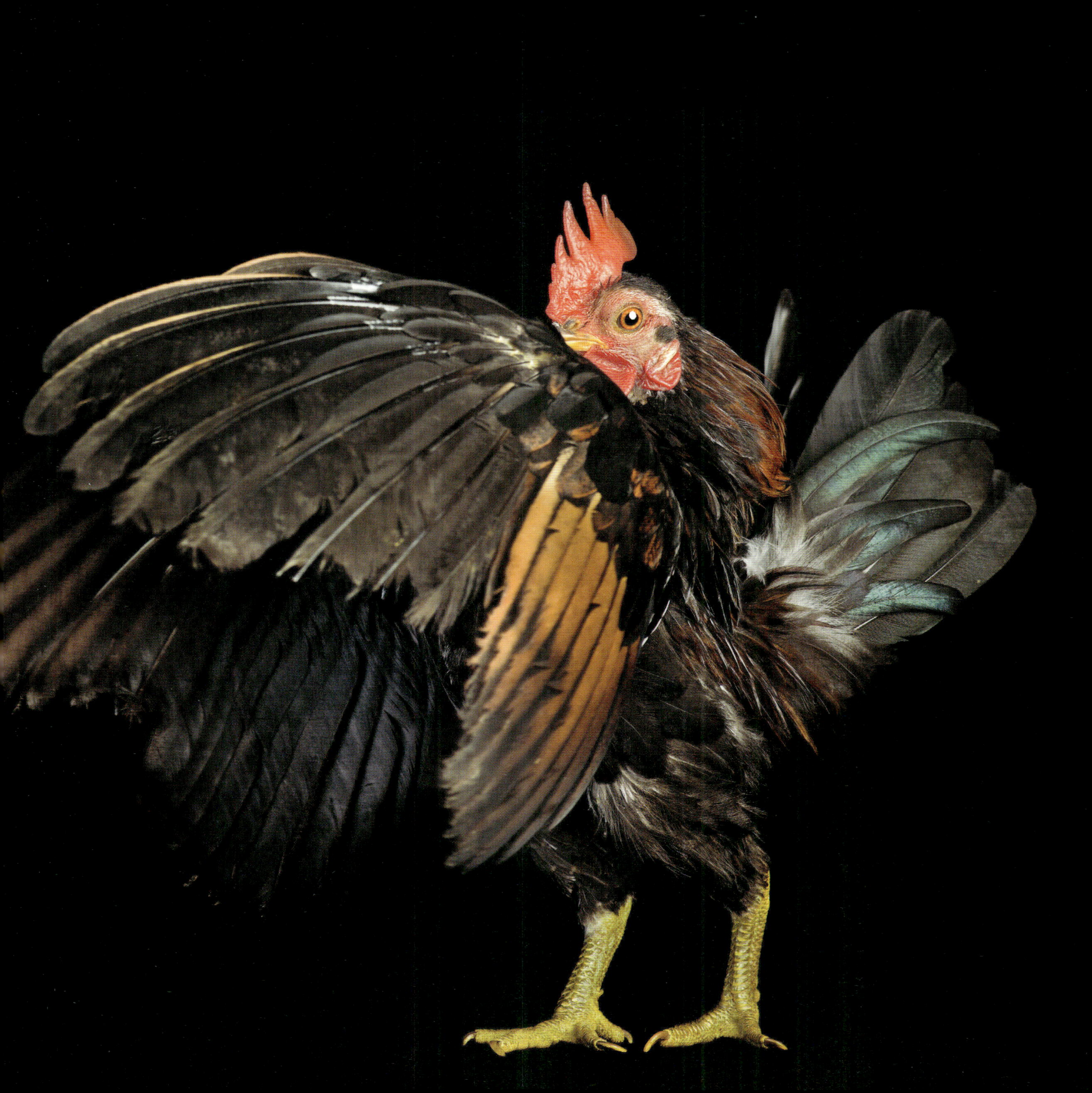

ANGRY

BIRDS

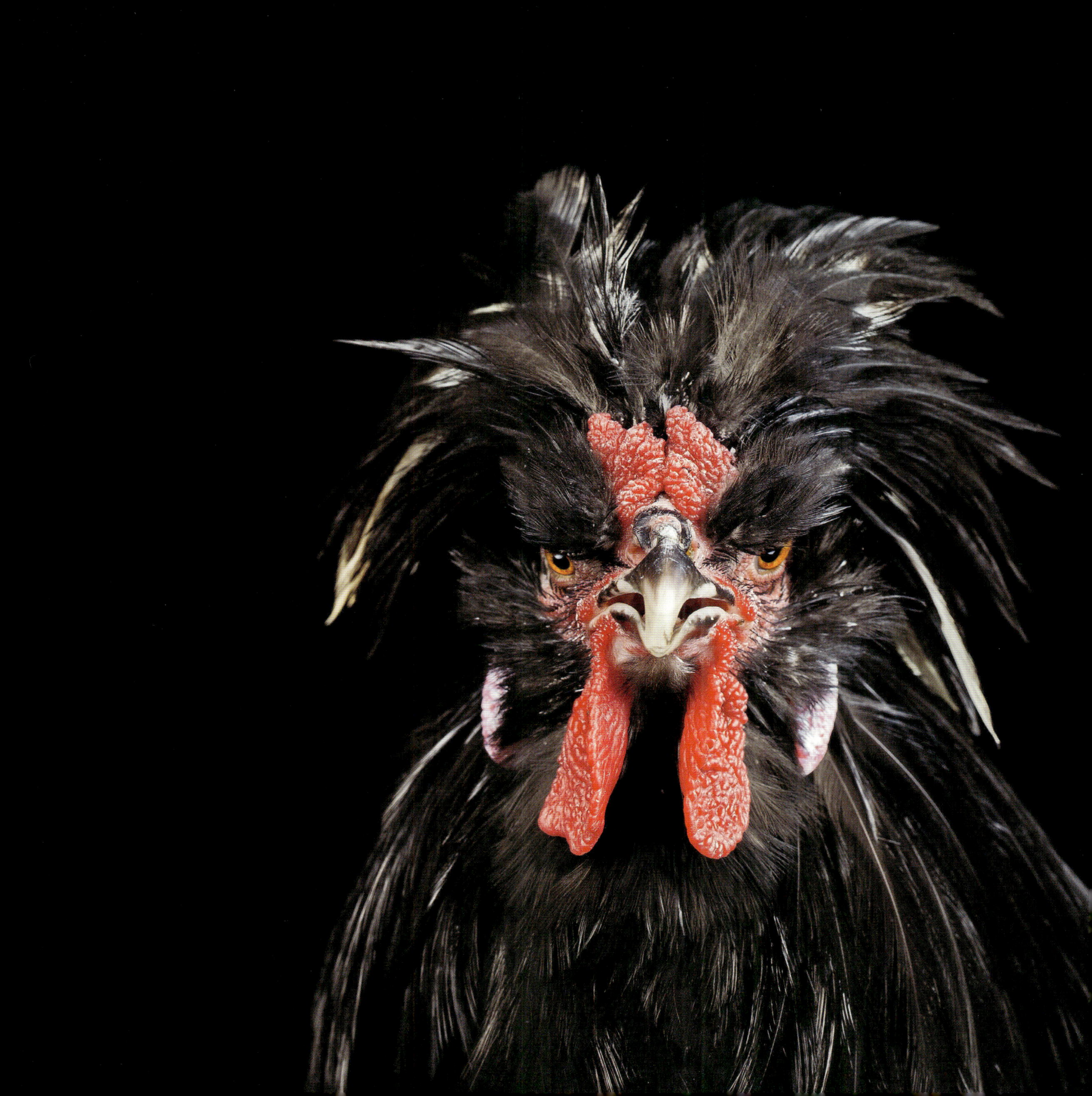

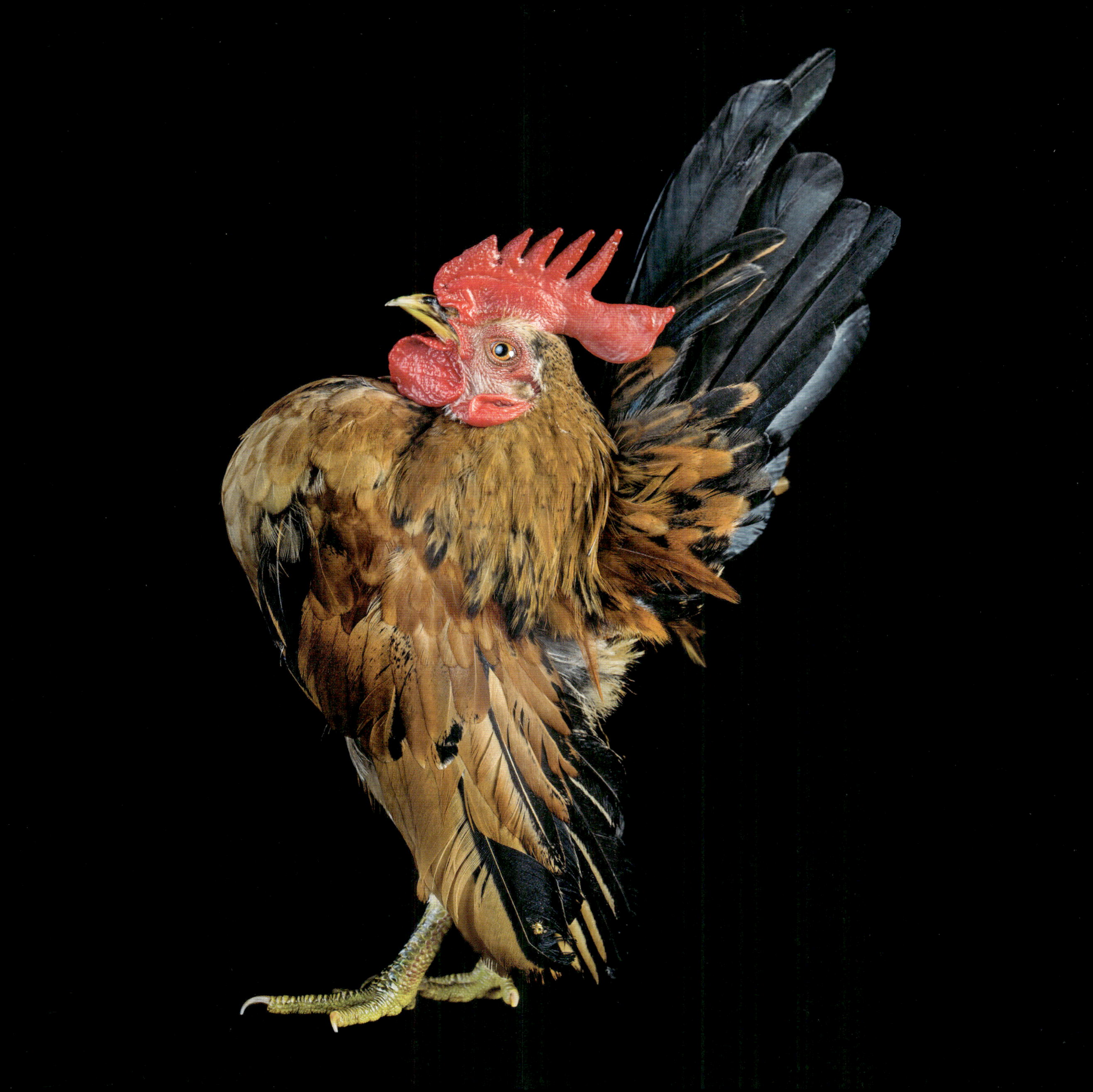

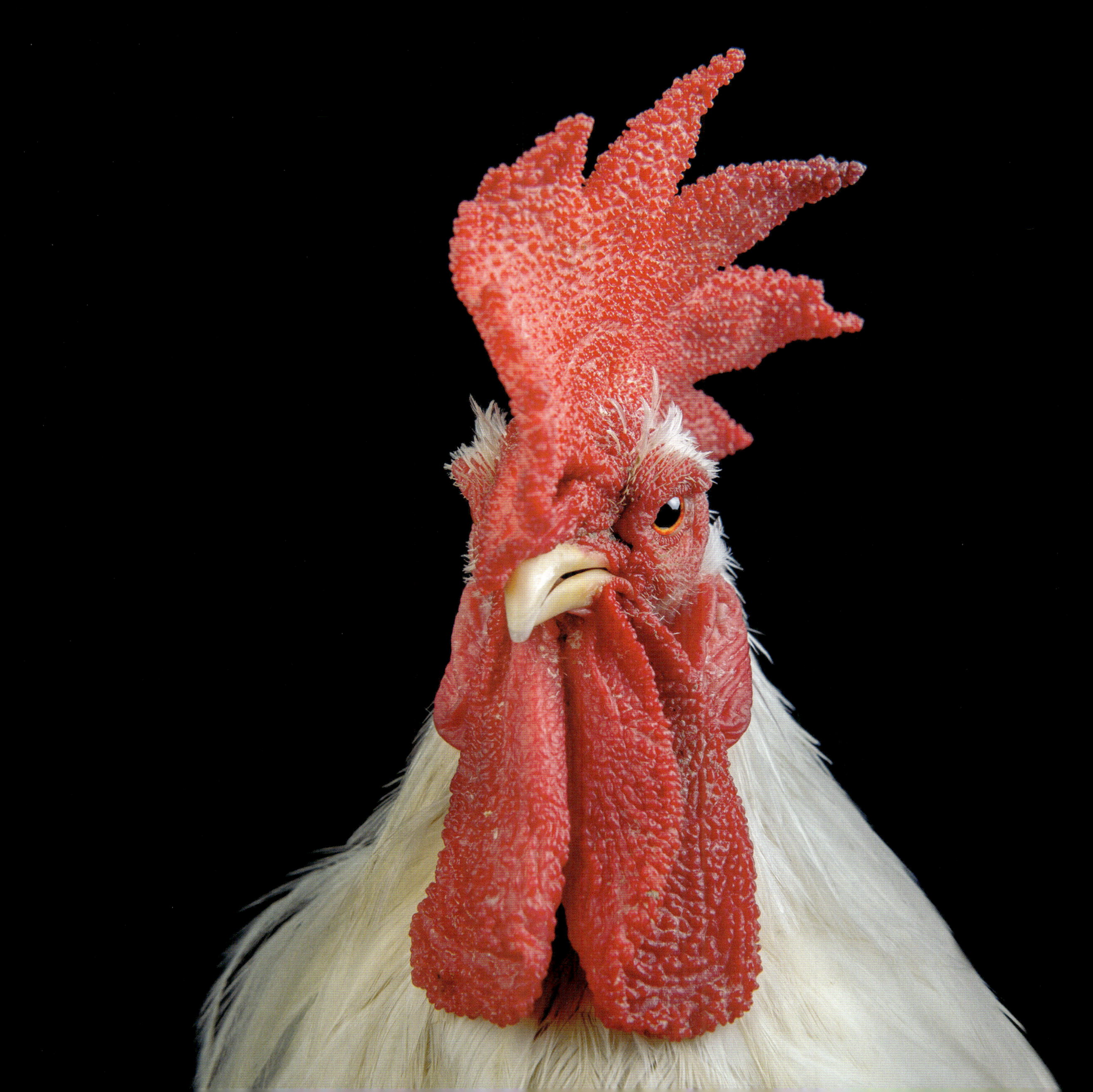

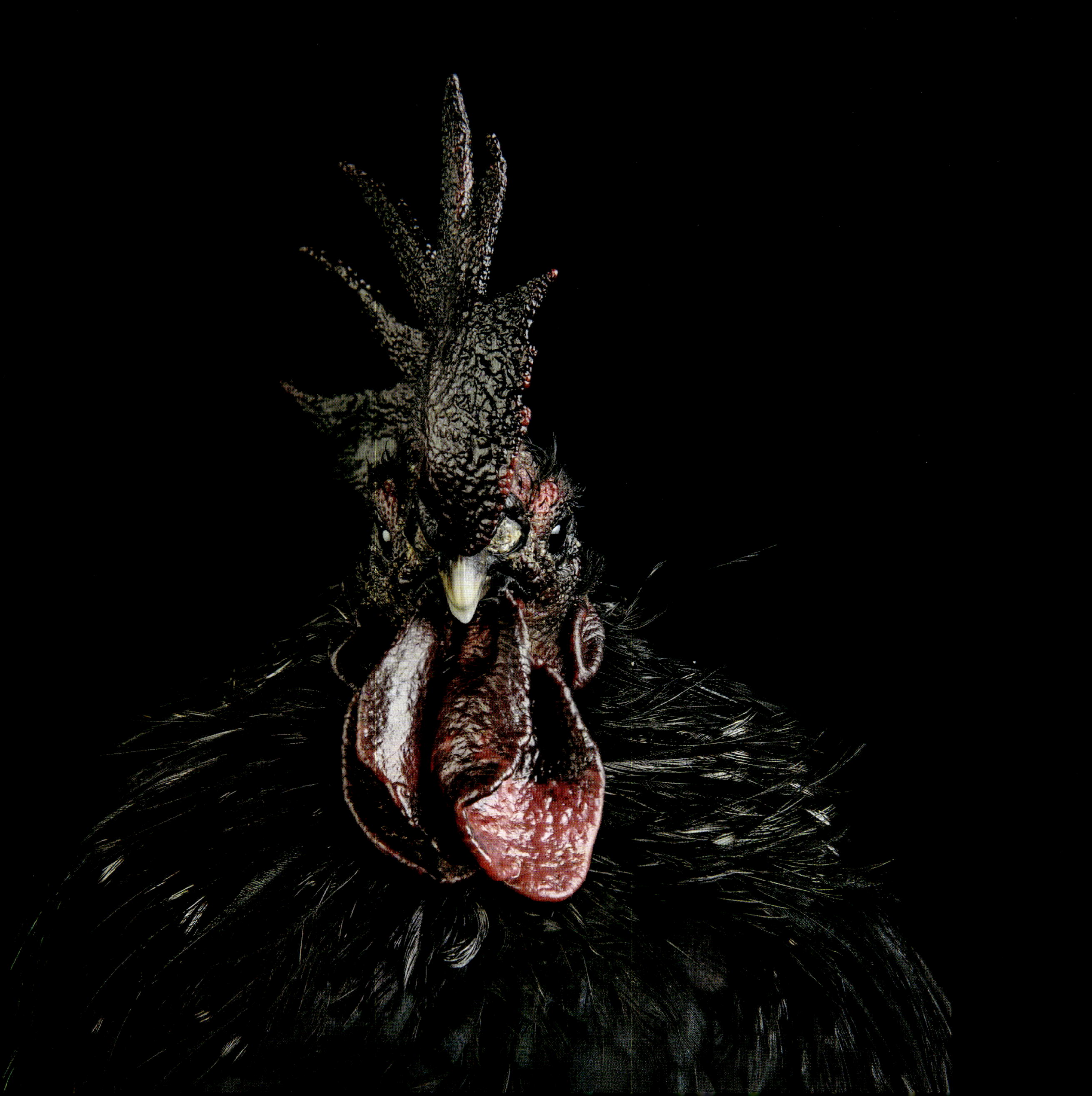

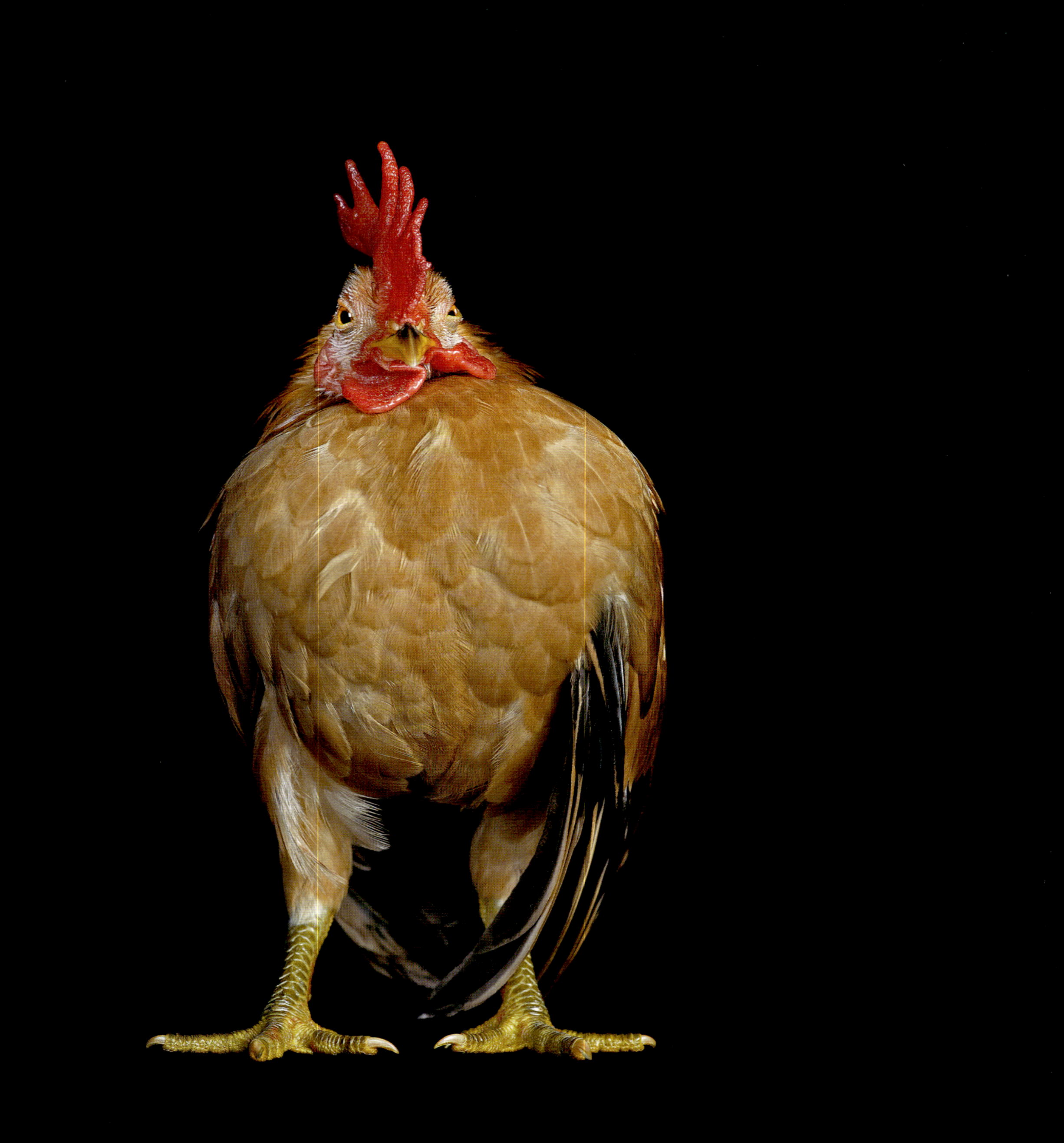

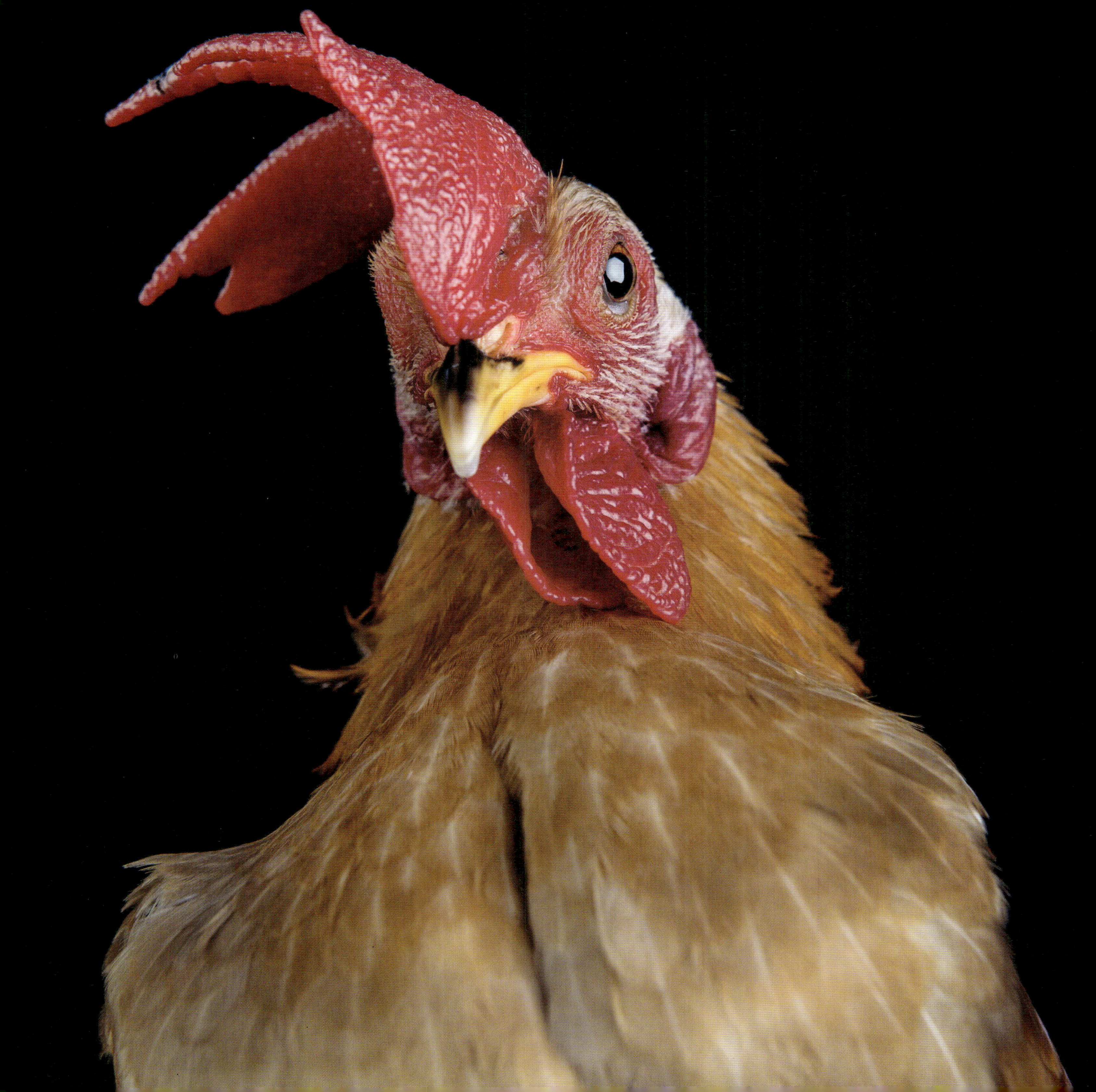

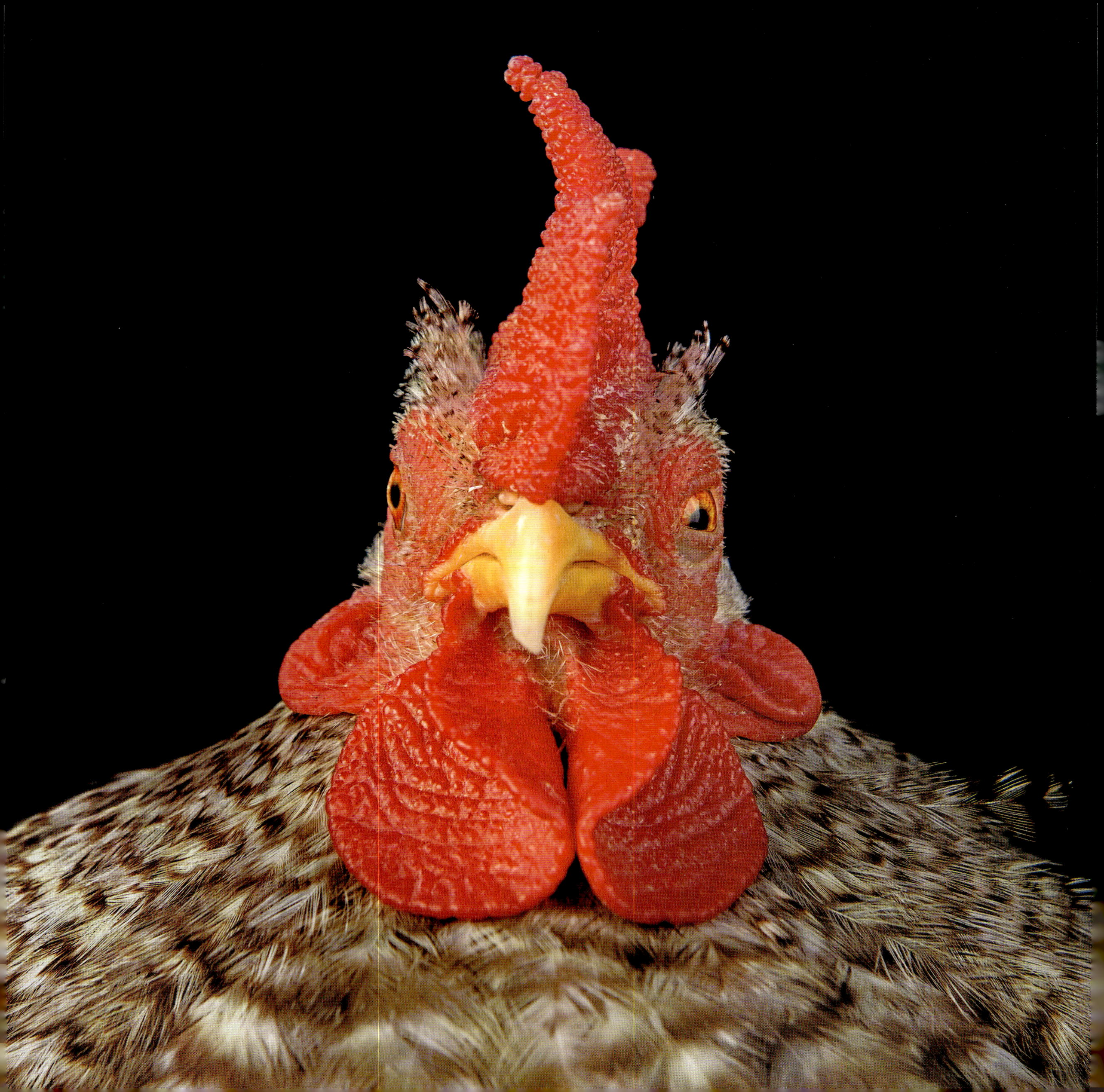

IF YOU CAN LOOK A
CHICKEN IN THE EYES

FOR TEN SECONDS,
IT'S YOURS FOREVER.

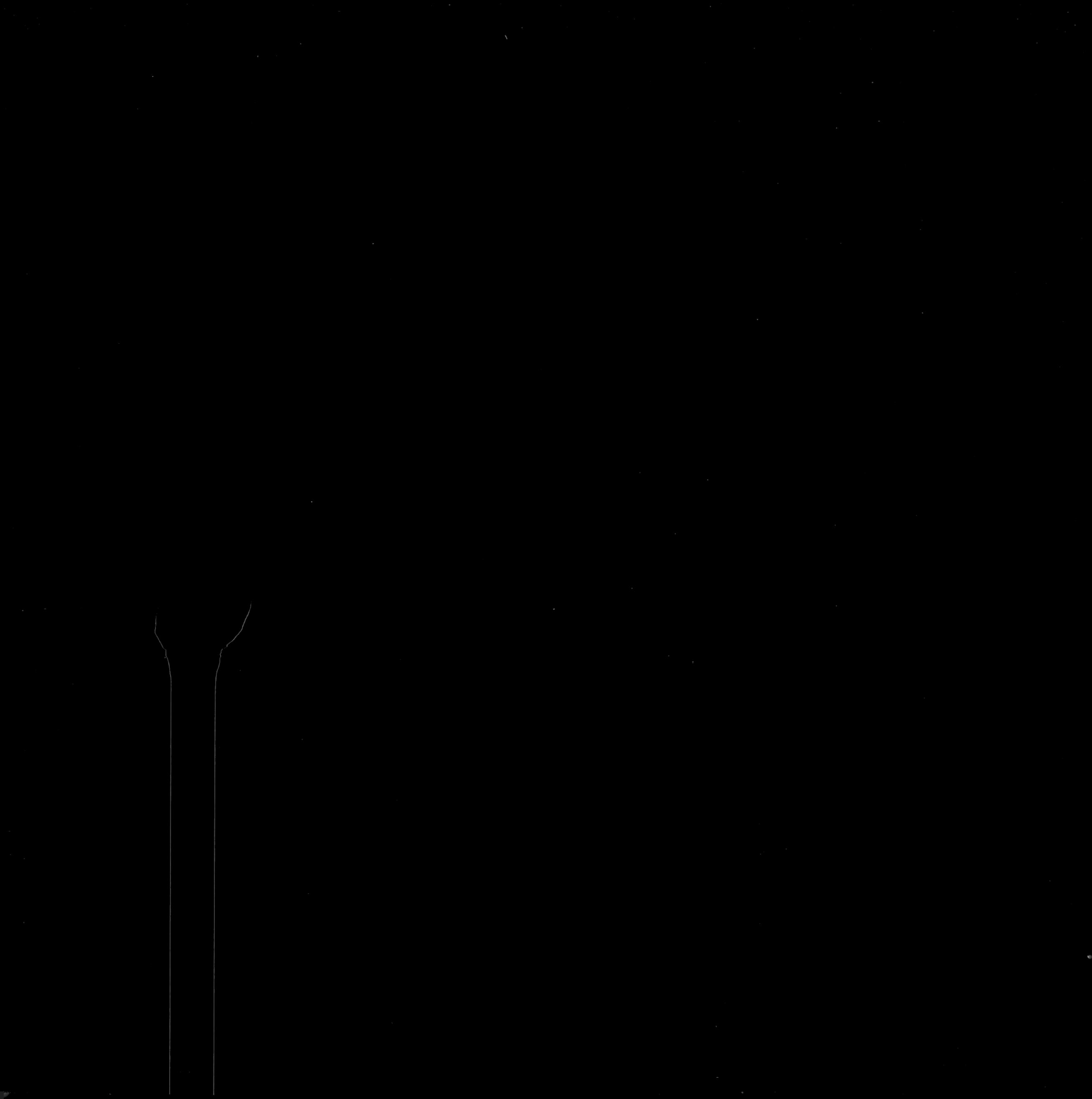

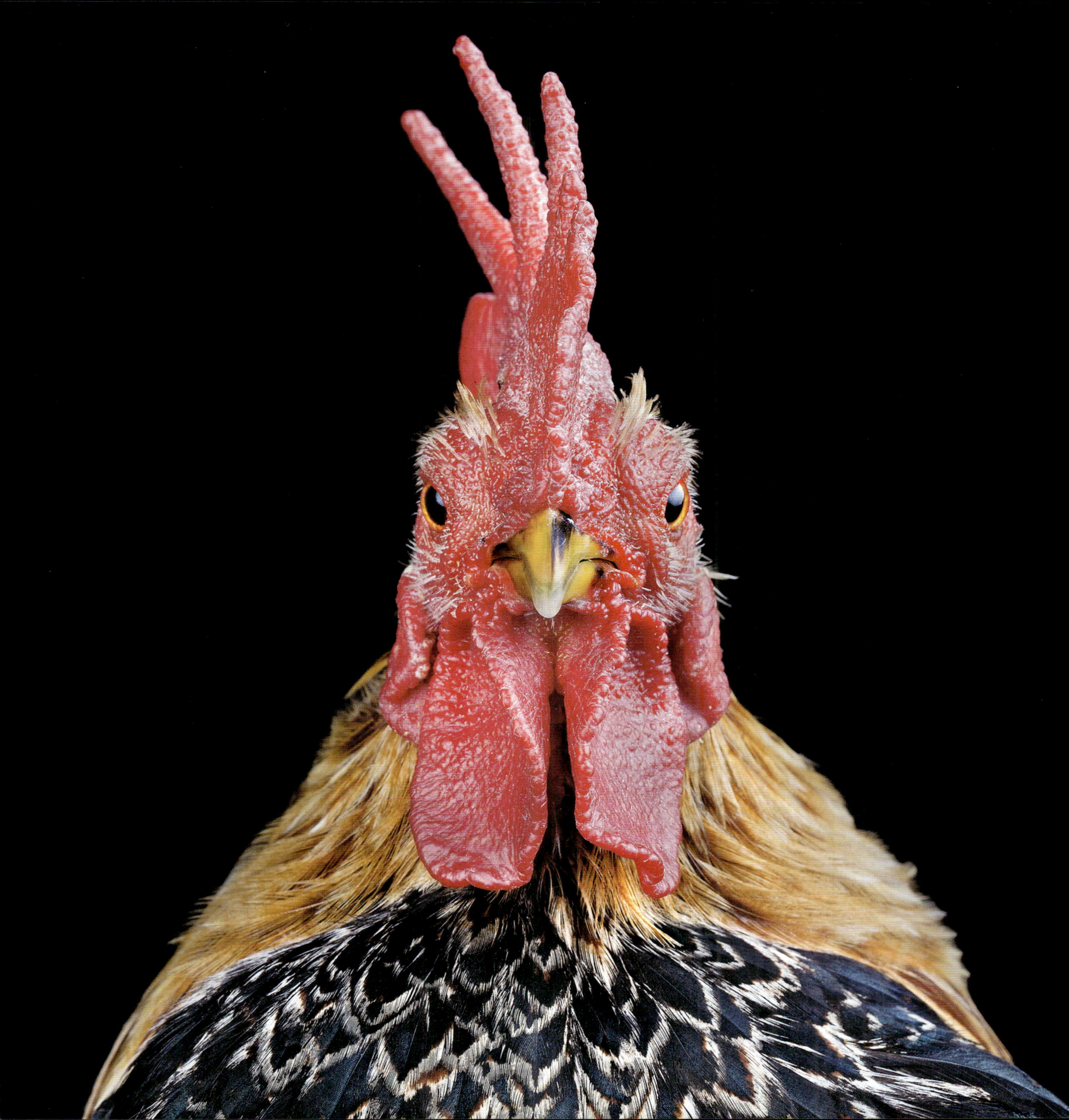

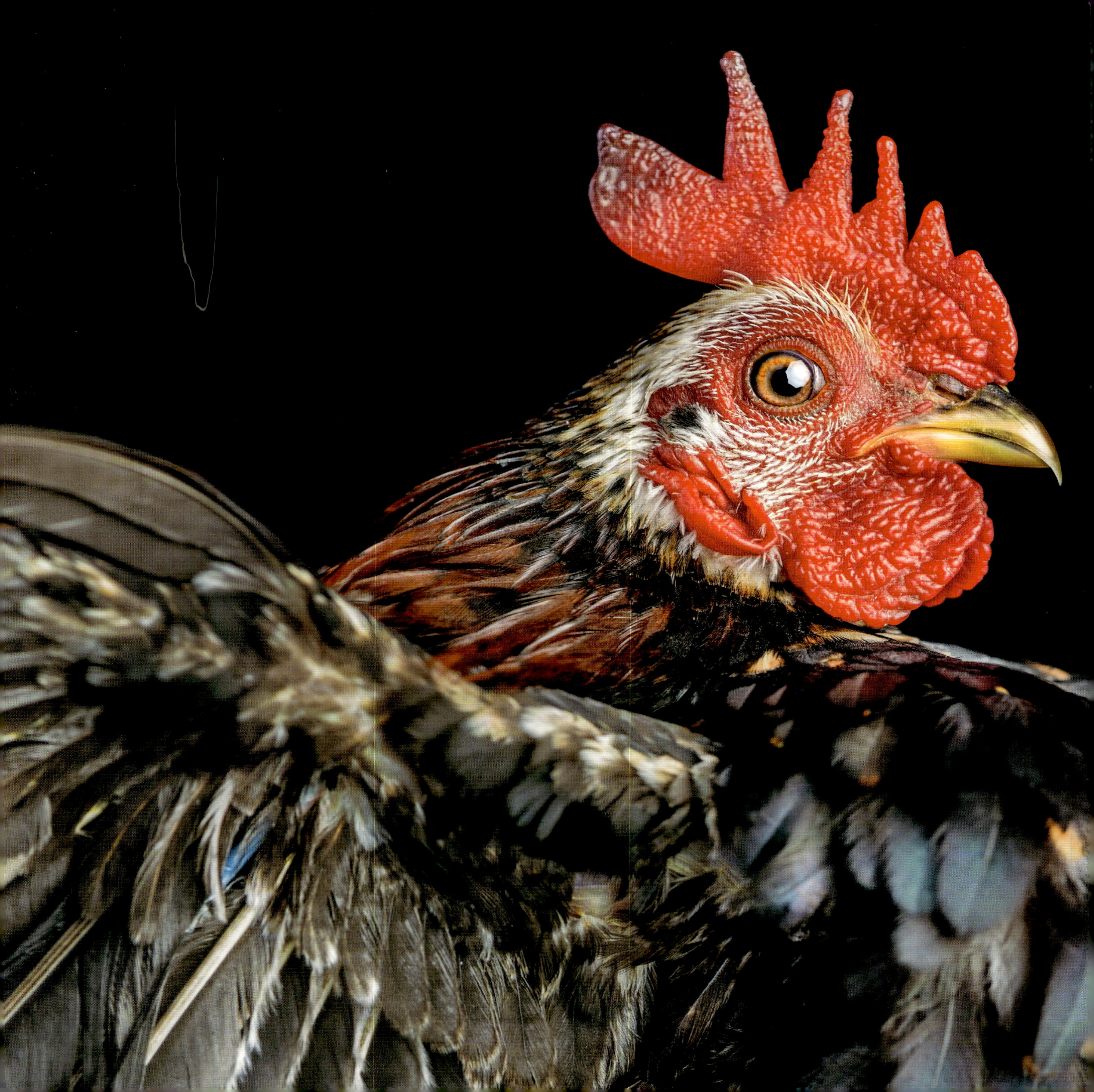

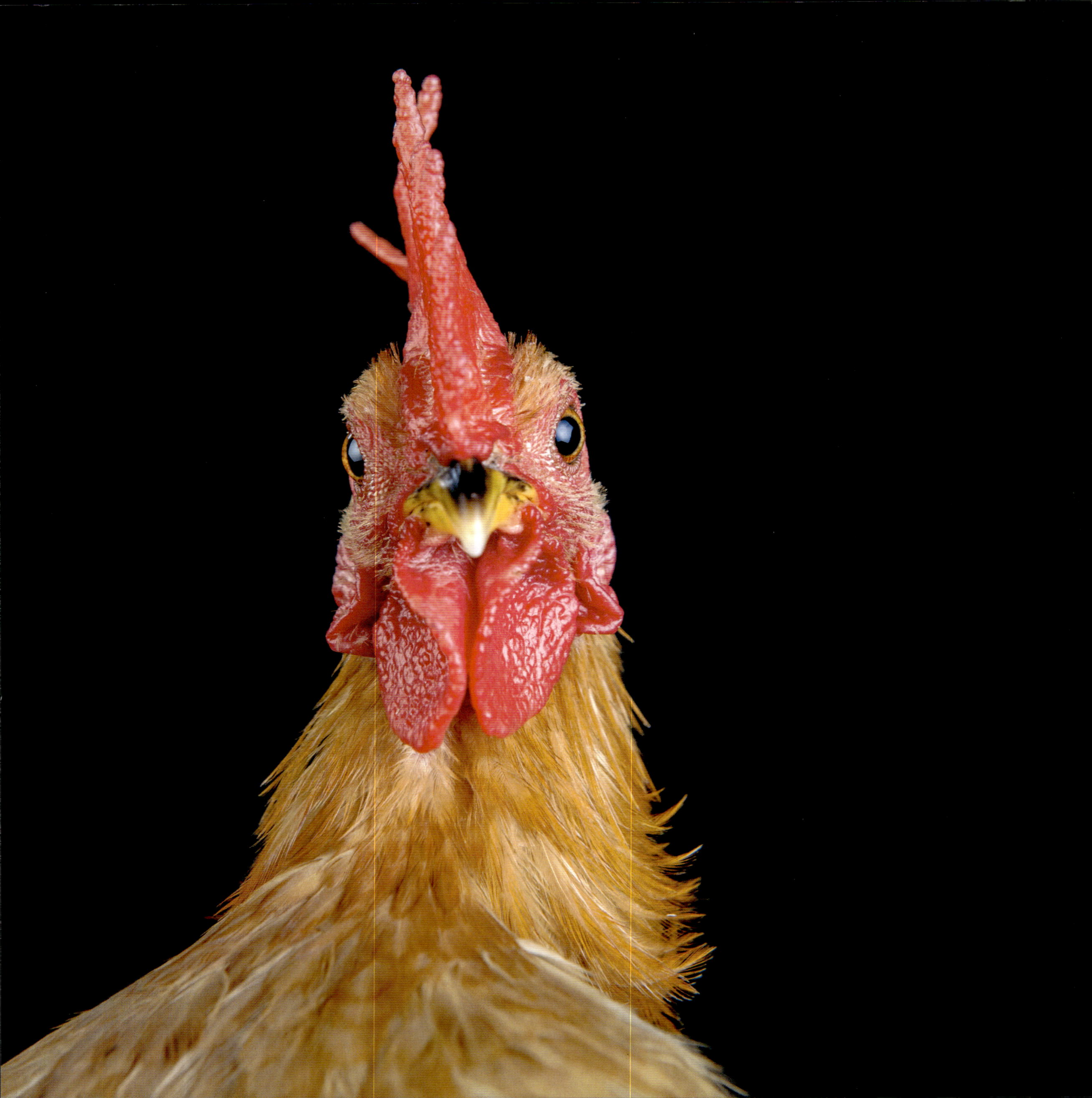

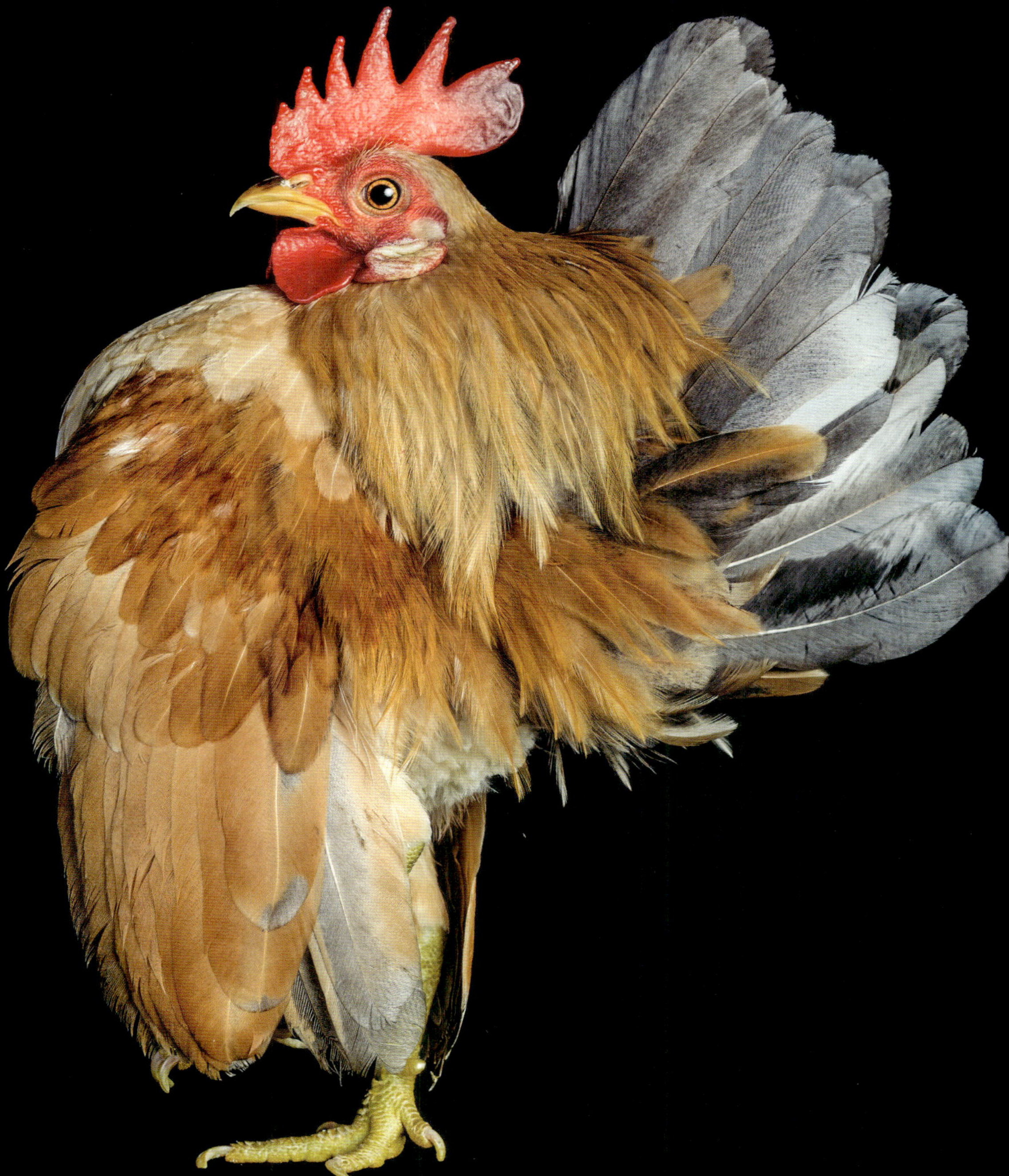

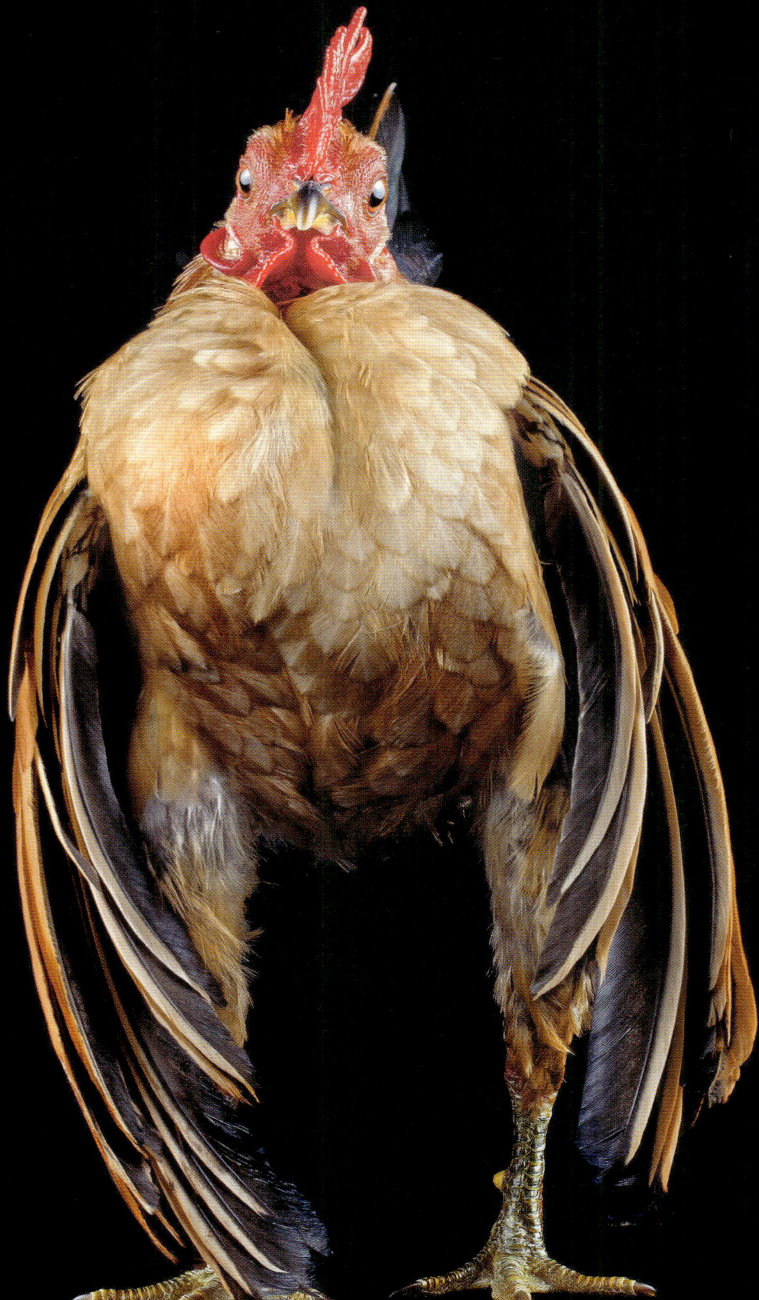

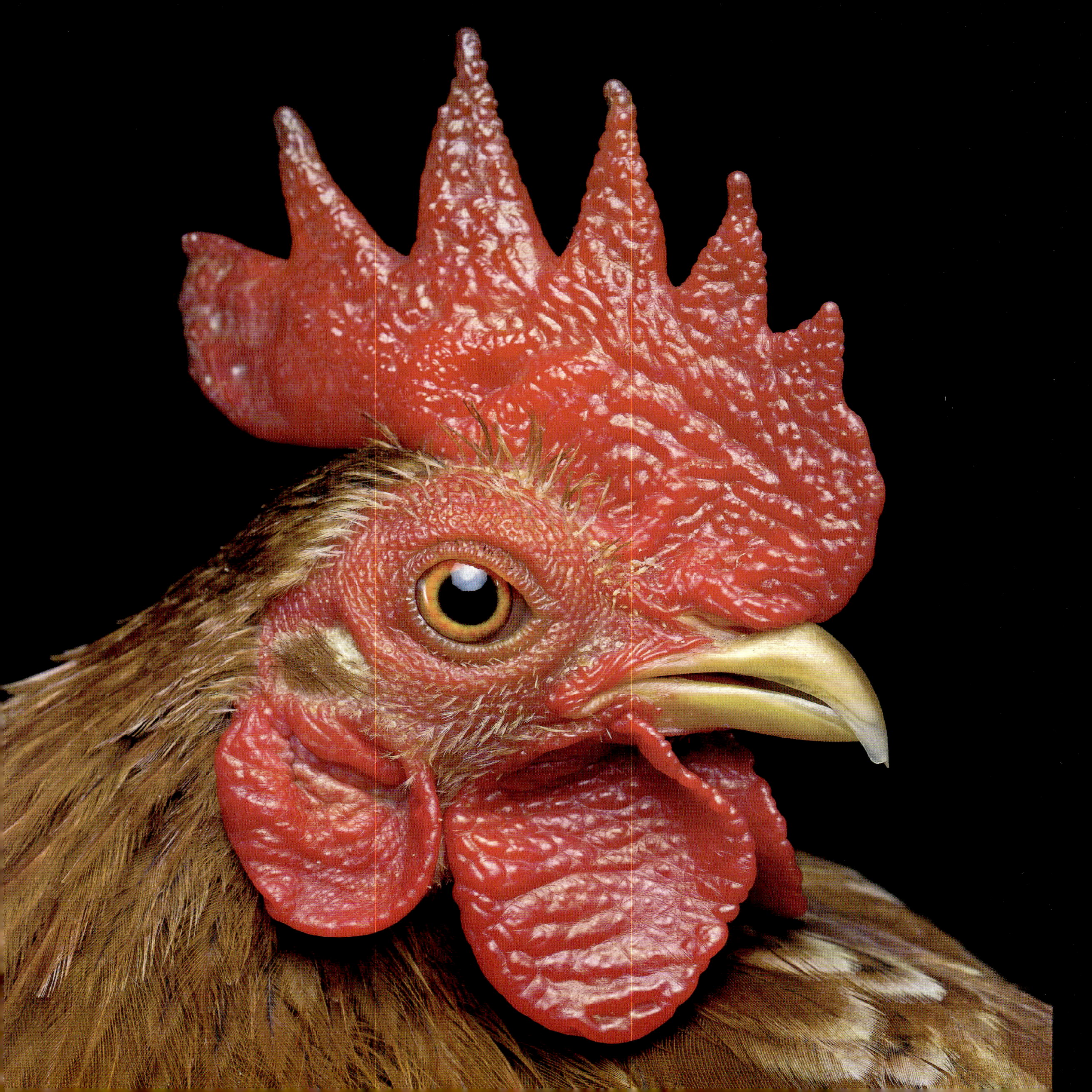

THE PHOTOGRAPHS

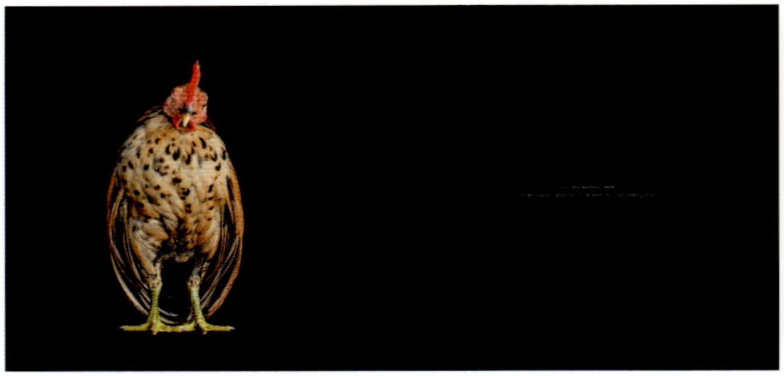

Serama (cock)

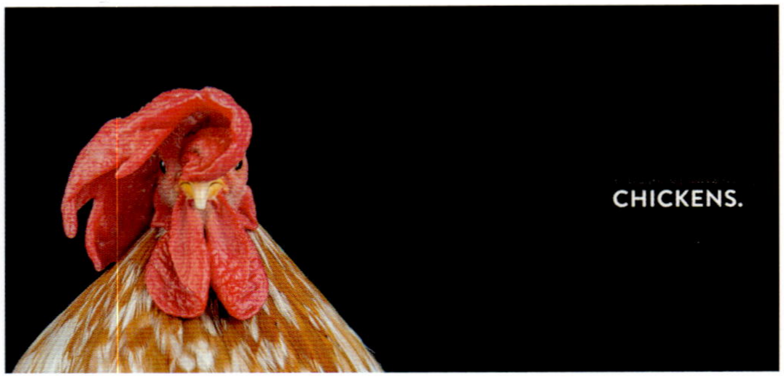

Japanese Bantam (cock)

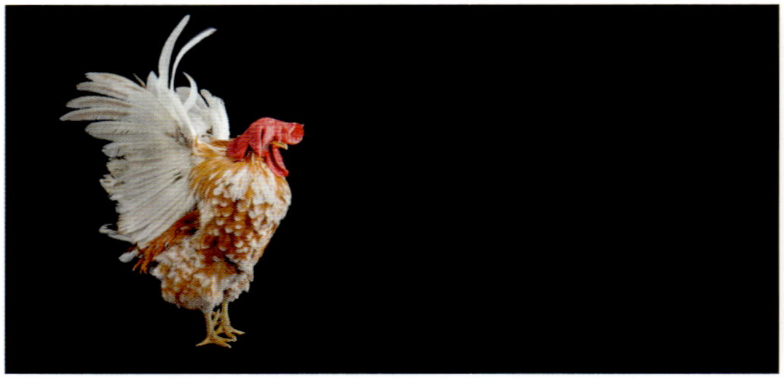

Japanese Bantam (cock)

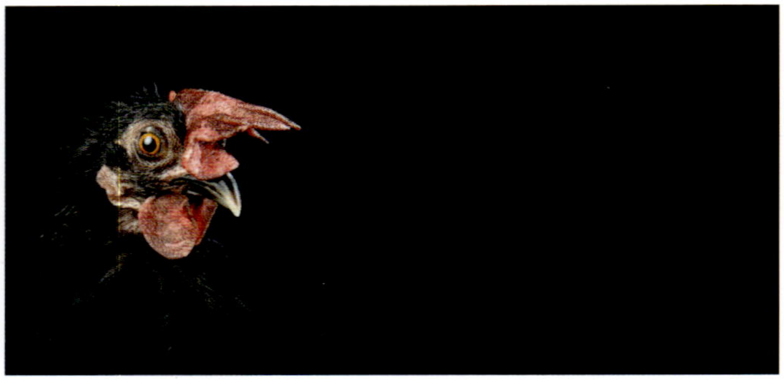

Japanese Bantam (hen)

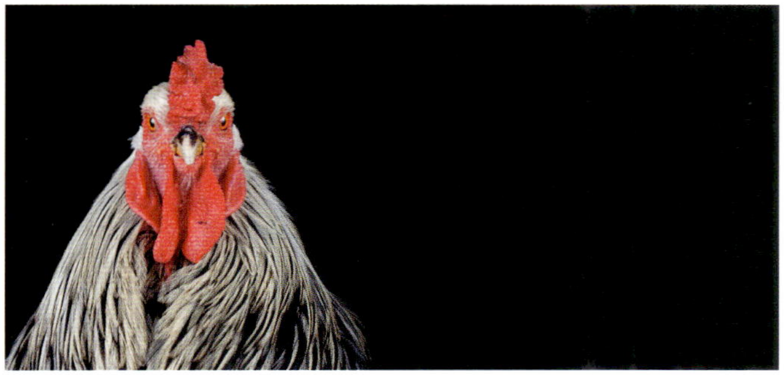

Light Brahma (cock)

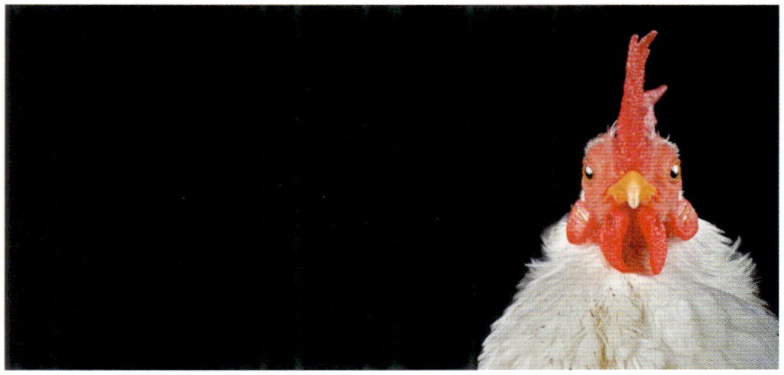

Serama (cock)

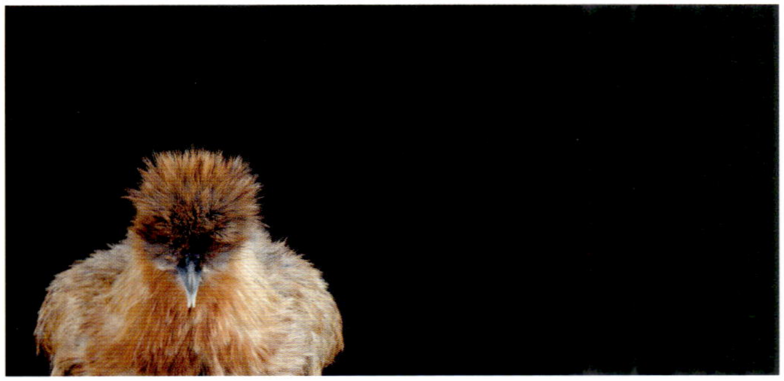

Brown Silkie (hen)

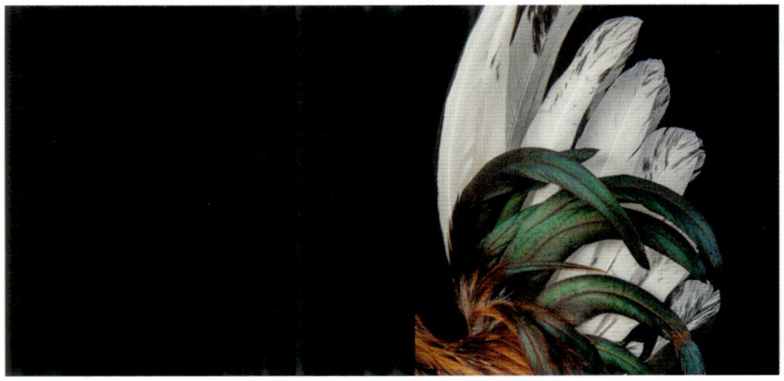

Serama (cock)

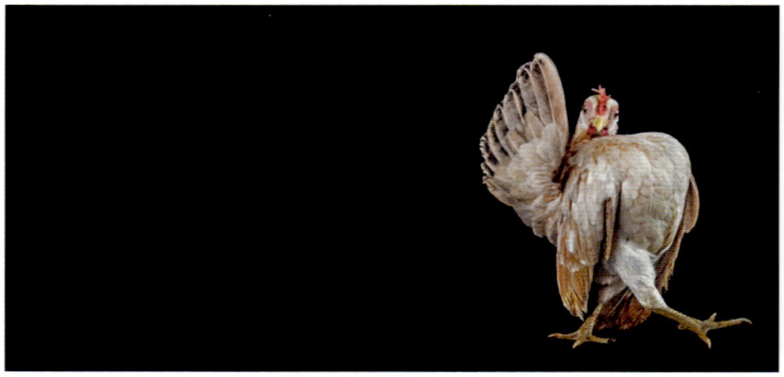

Serama (cock)

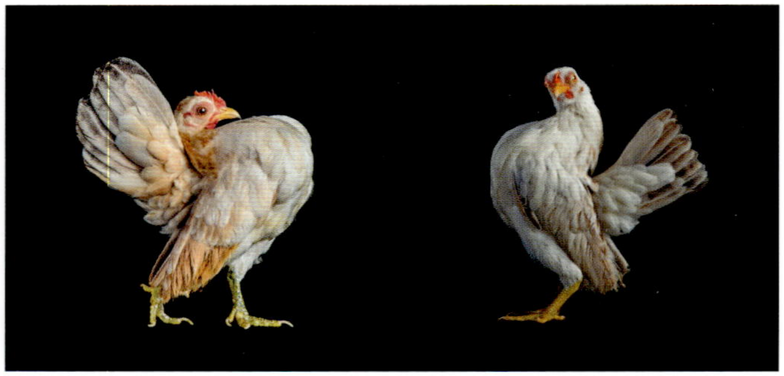

Serama (cock)

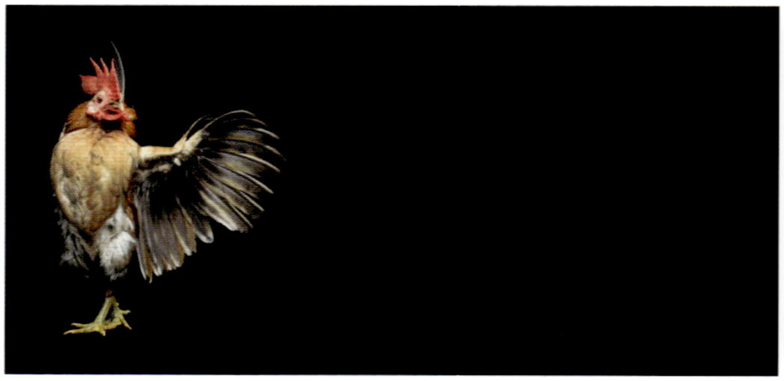

Serama (cock)

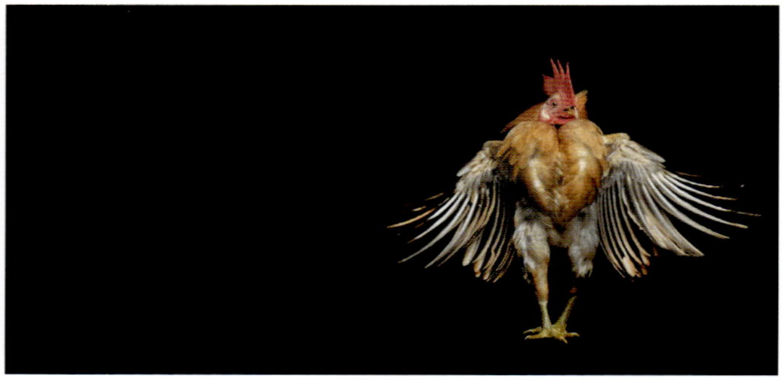

Serama (cock)

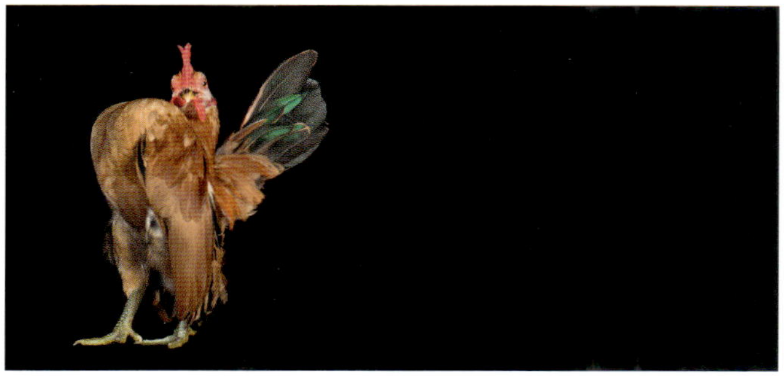

Serama (cock)

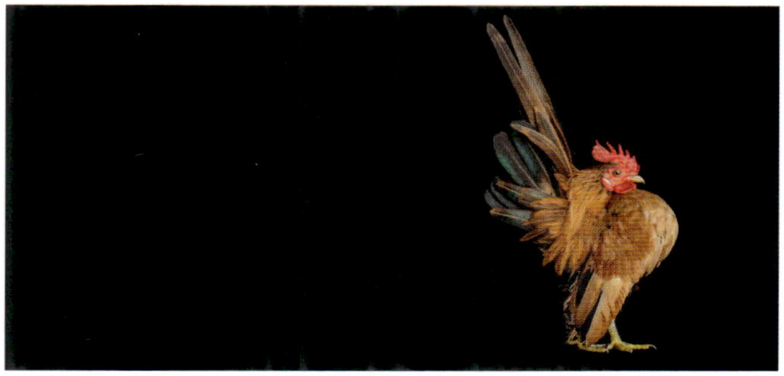

Serama (cock)

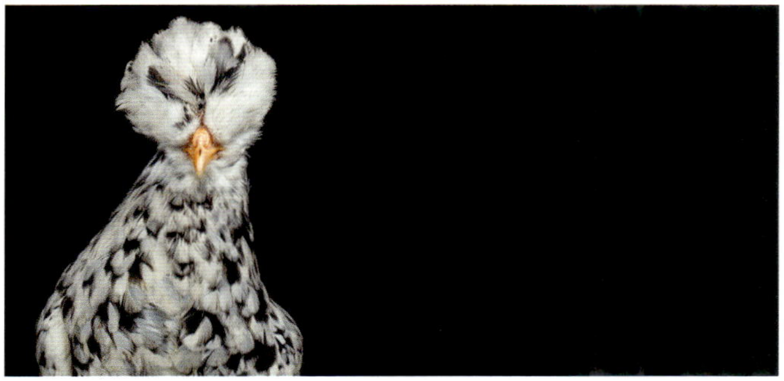

Polish (hen)

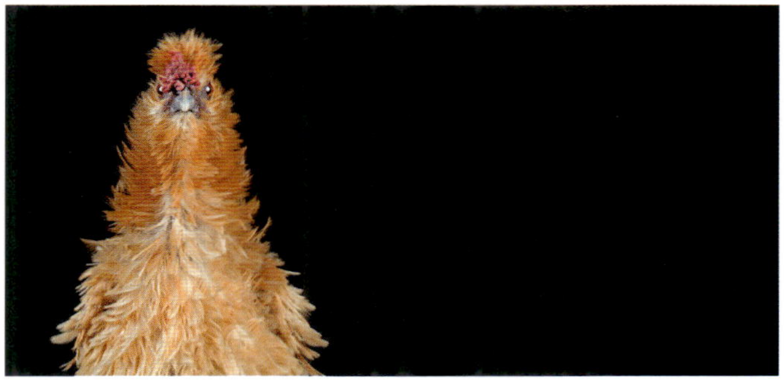

Brown Frizzle-Silkie (hen)

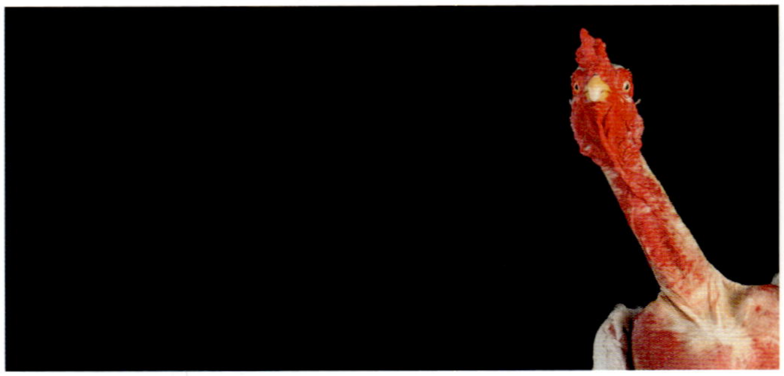

Thai Naked-neck (hen)

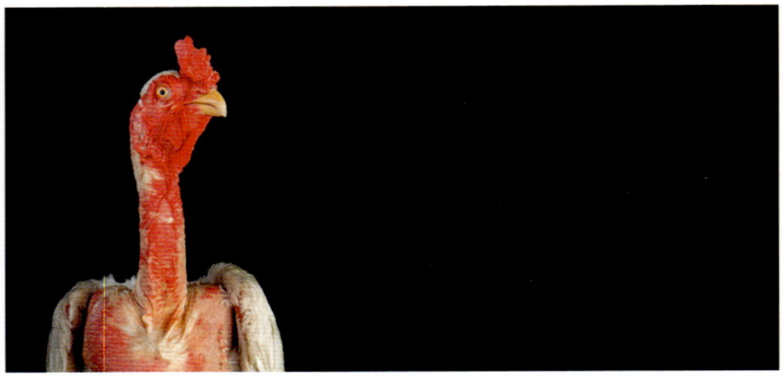

Thai Naked-neck (hen)

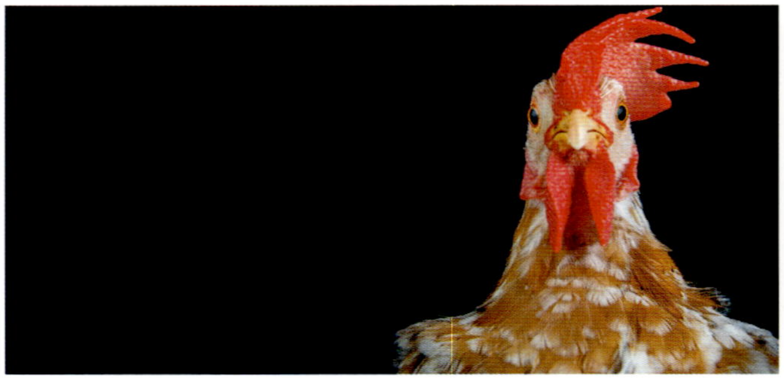

Serama (hen)

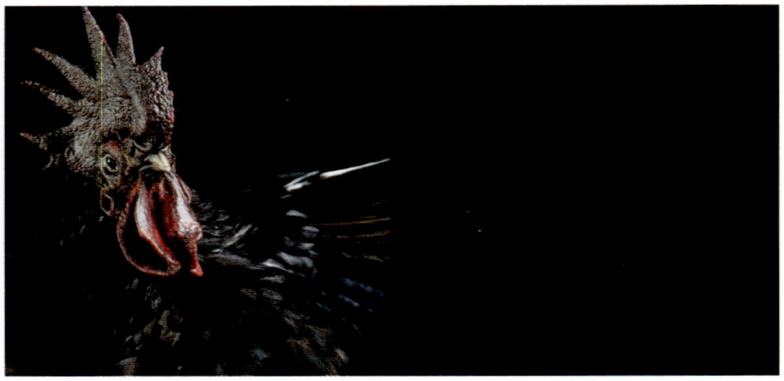

Japanese Bantam (cock)

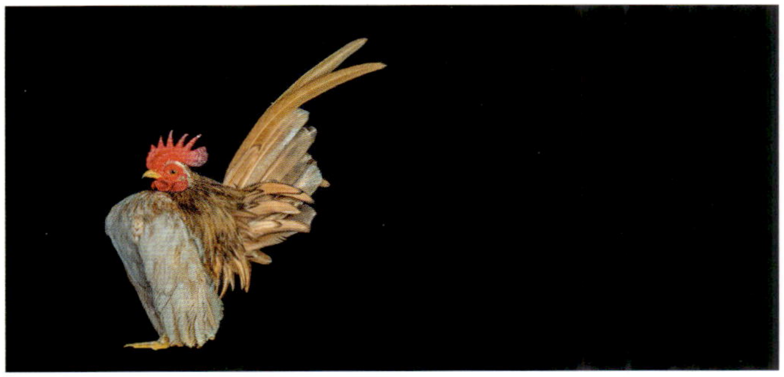

Serama (cock)

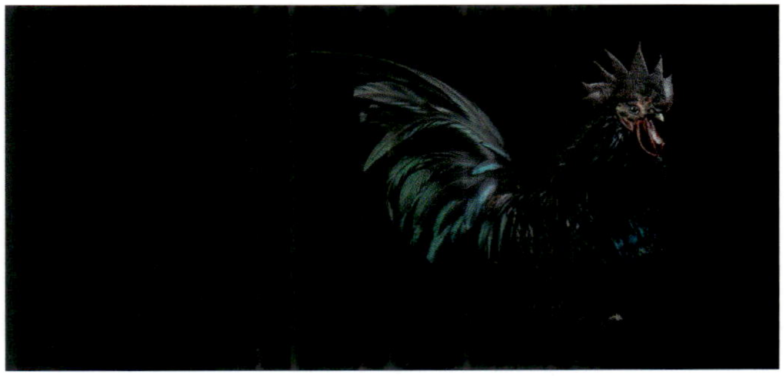

Japanese Bantam (cock)

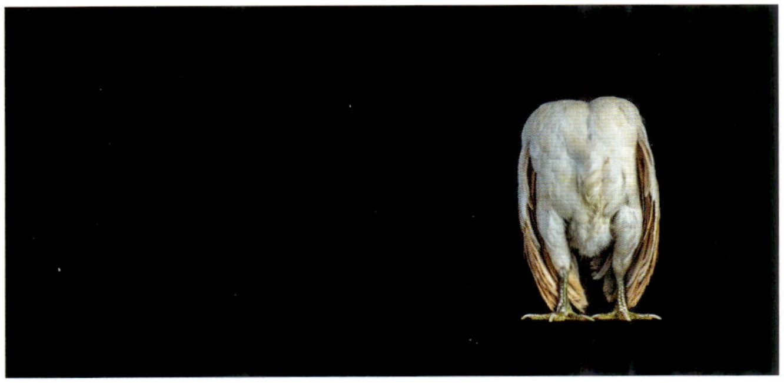

Serama (cock)

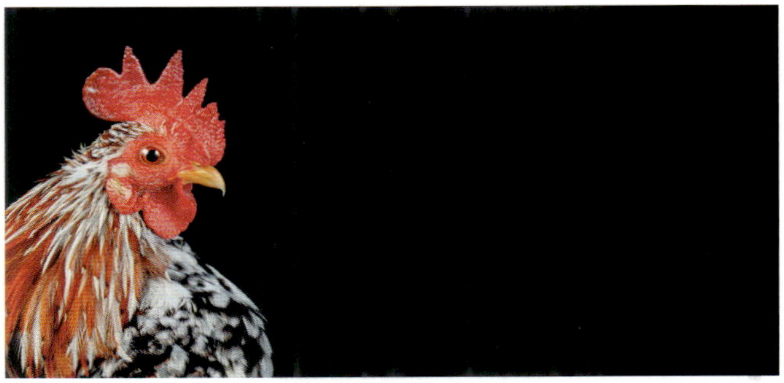

Serama (cock)

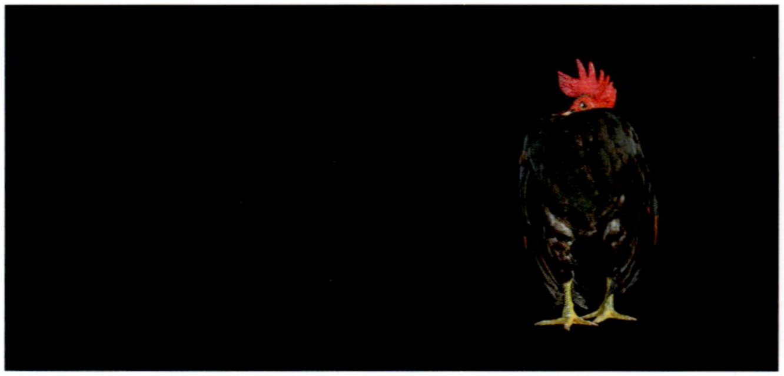

Serama (cock)

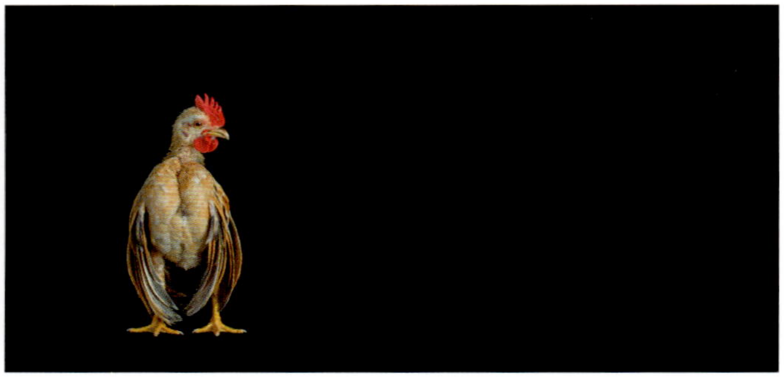

Serama (cock)

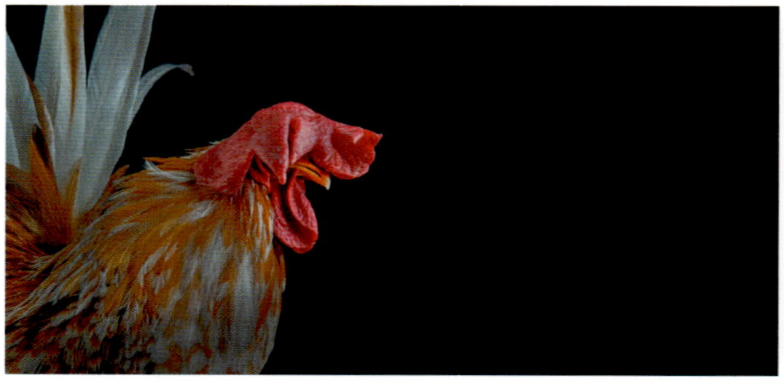

Japanese Bantam (cock)

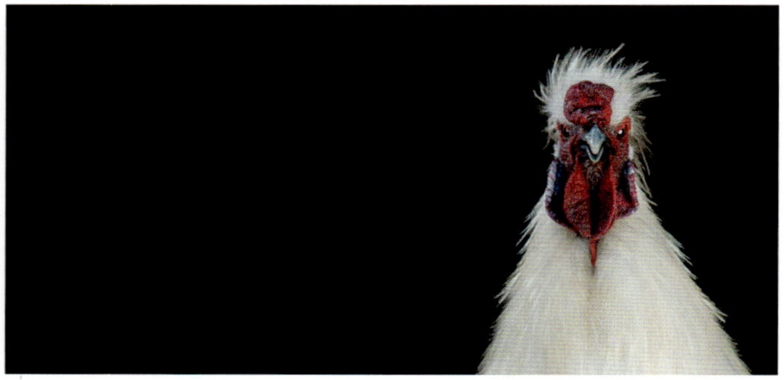

Silkie (cock)

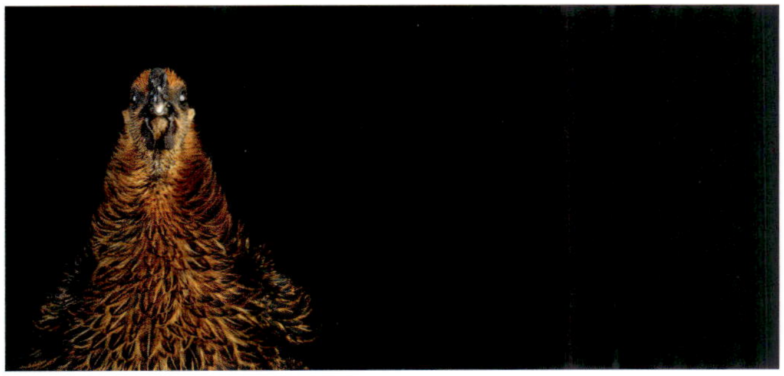

Frizzle Bantam (hen)

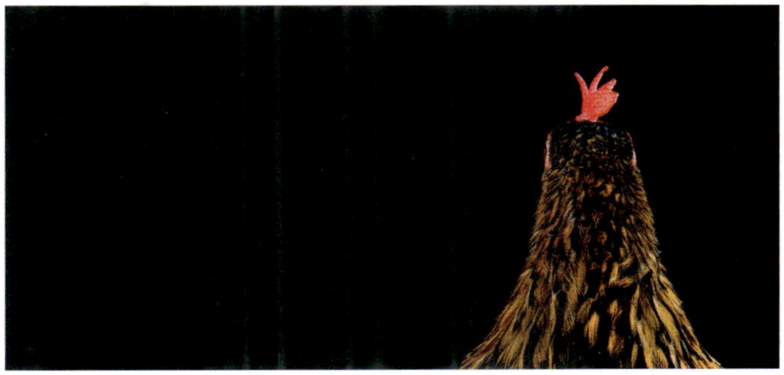

Ayam Serama (hen)

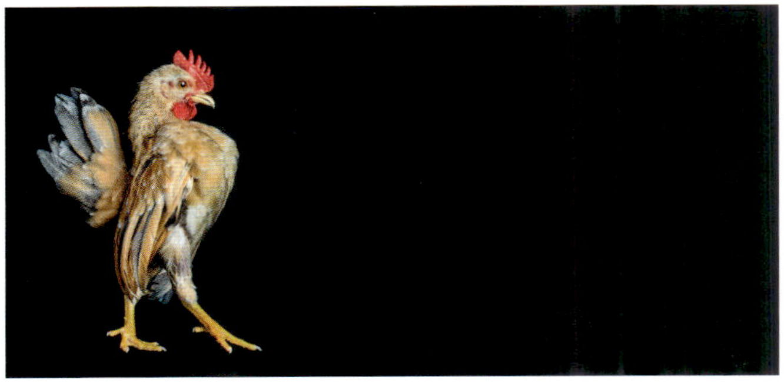

Serama (cock)

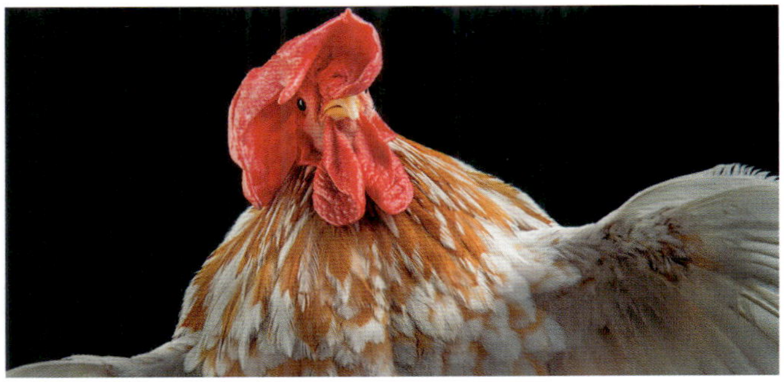

Japanese Bantam (cock)

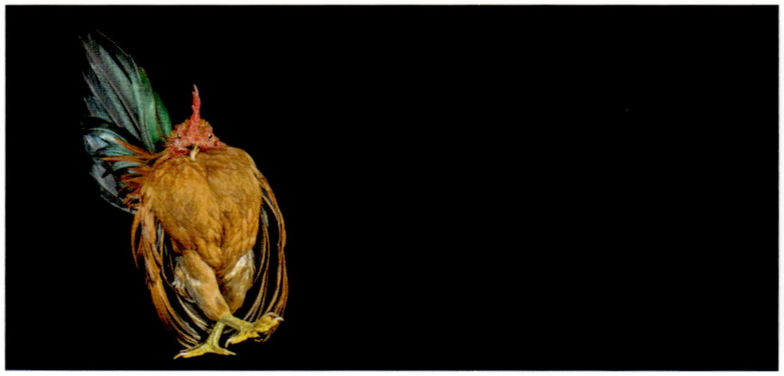

Serama (cock)

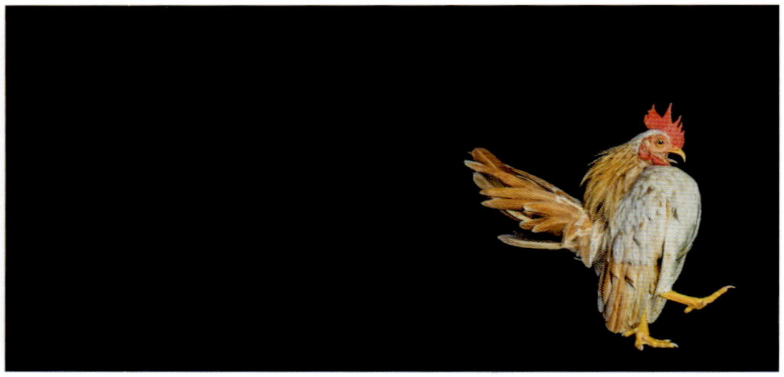

Serama (cock)

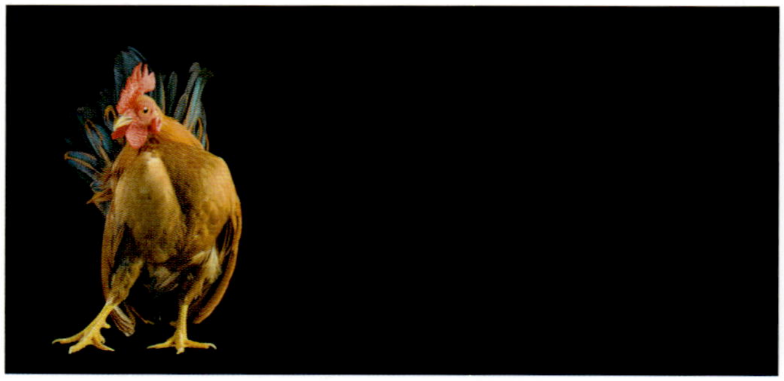

Serama (cock)

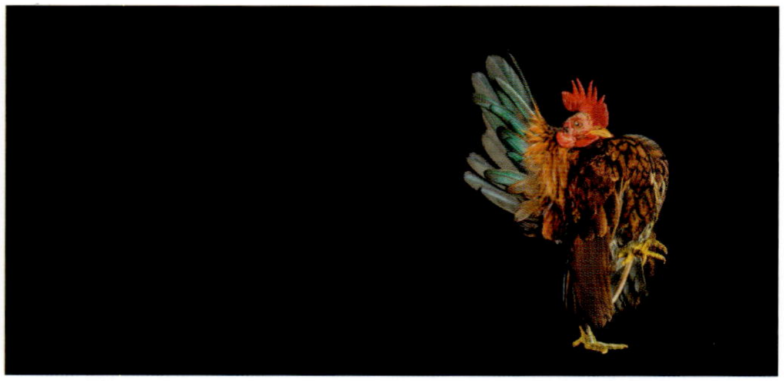

Serama (cock)

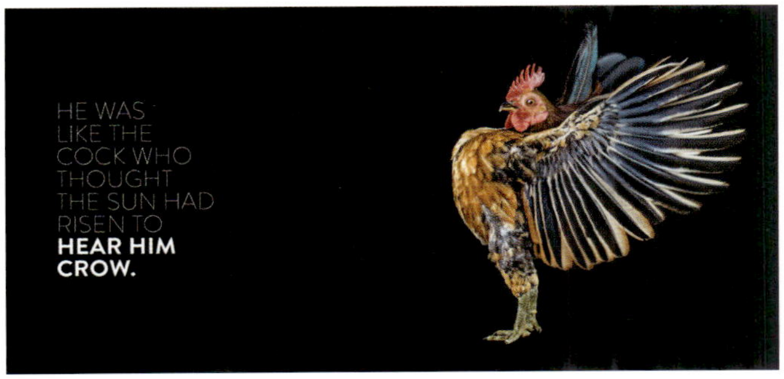

Serama (cock)

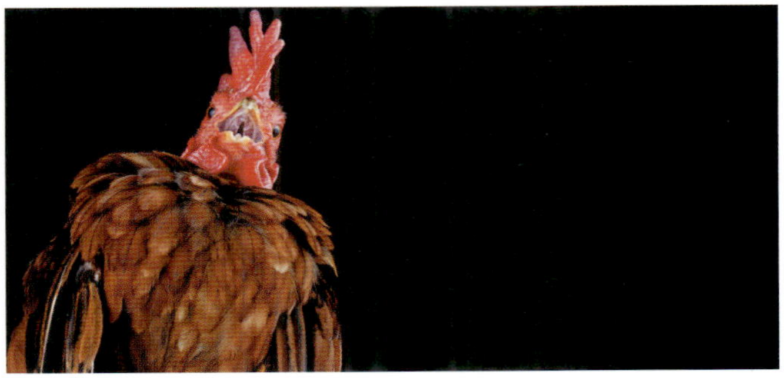

Serama (cock)

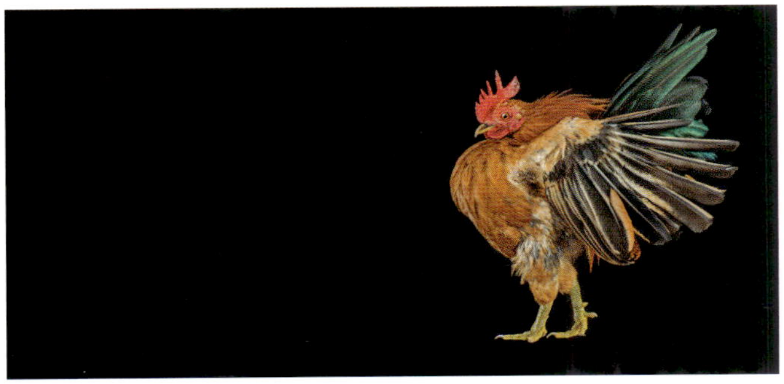

Serama (cock)

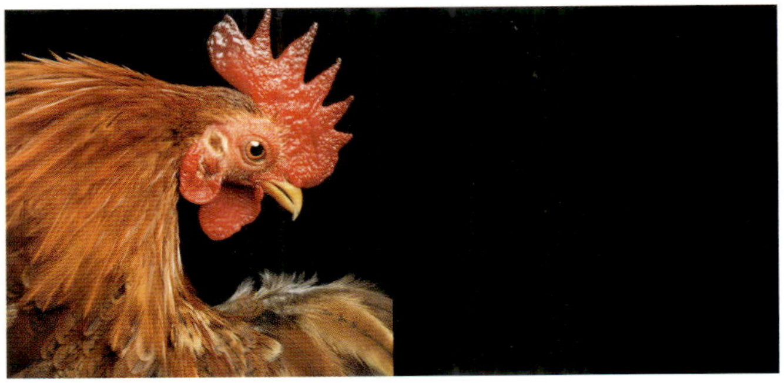

Serama (cock)

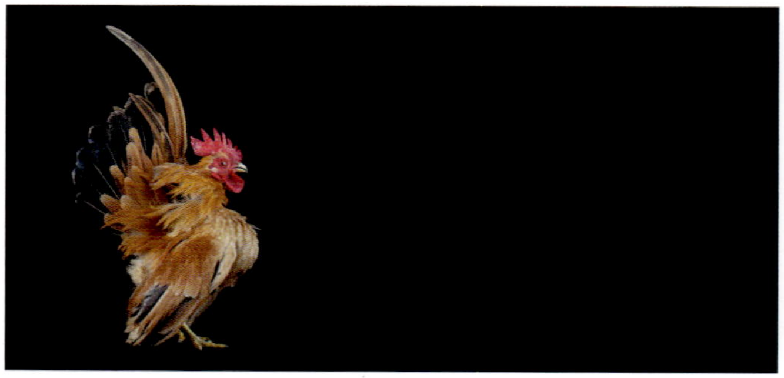

Serama (cock)

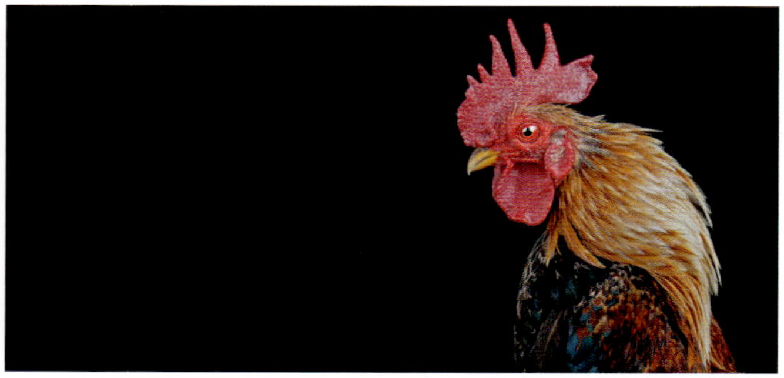

Serama (cock)

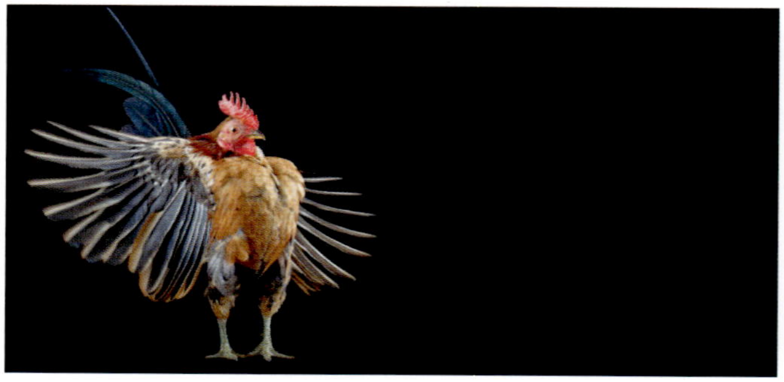

Serama (cock)

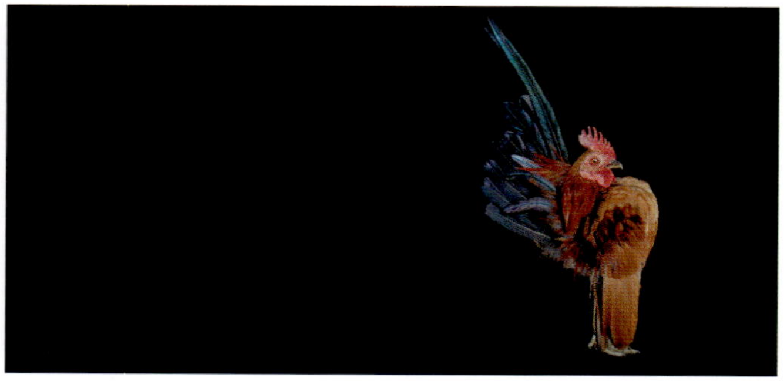

Serama (cock)

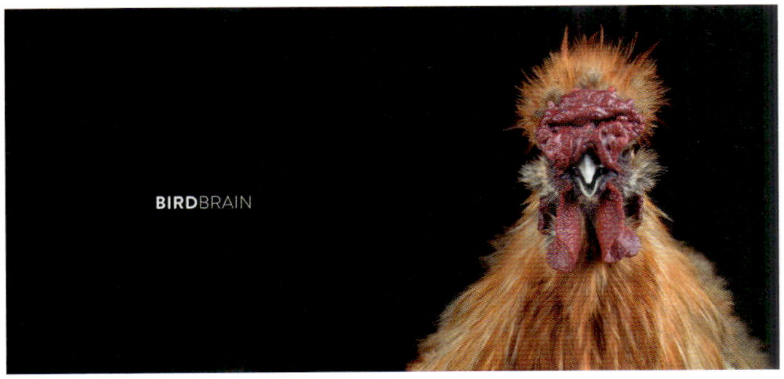

Brown Silkie (hen)

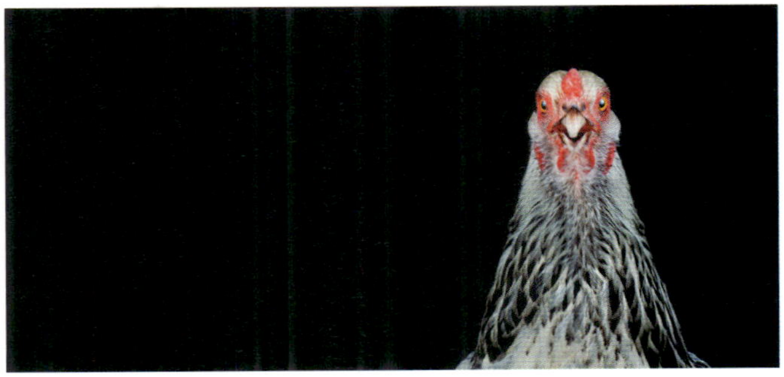

Light Brahma (hen)

White Silkie (hen)

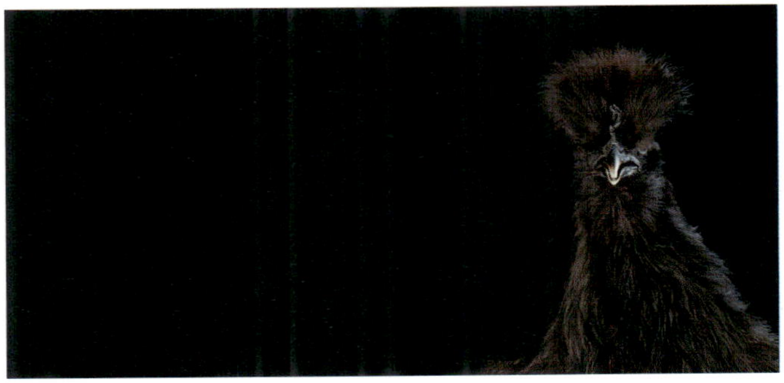

Black Silkie (hen)

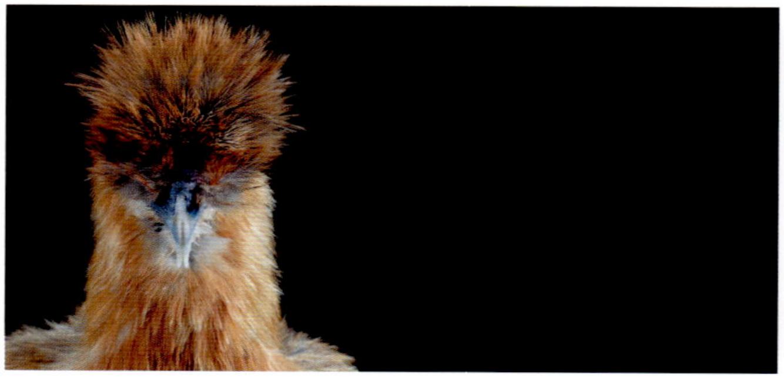

Brown Silkie (hen)

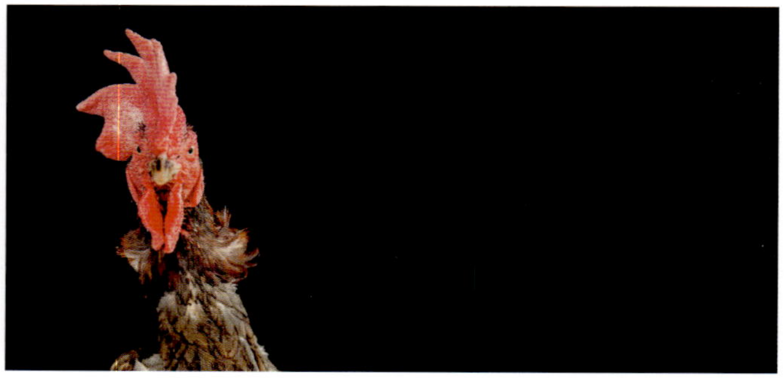

Thai game fowl (cock)

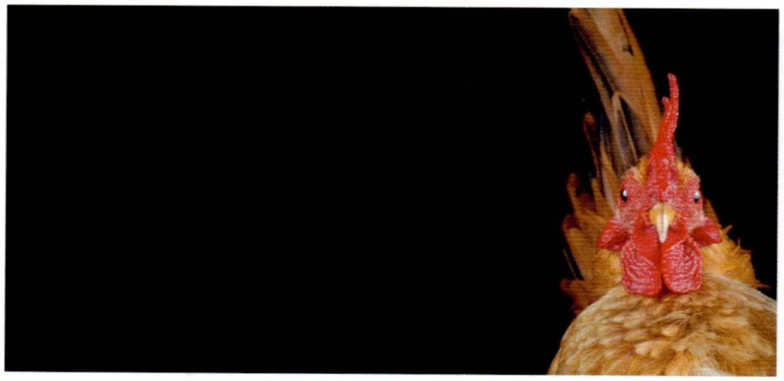

Serama (cock)

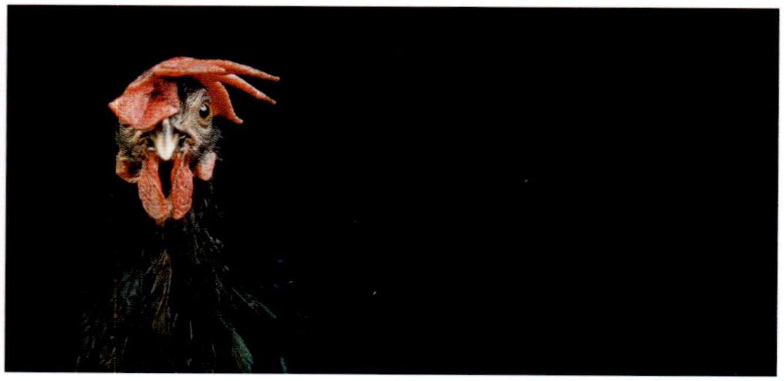

Japanese Bantam (hen)

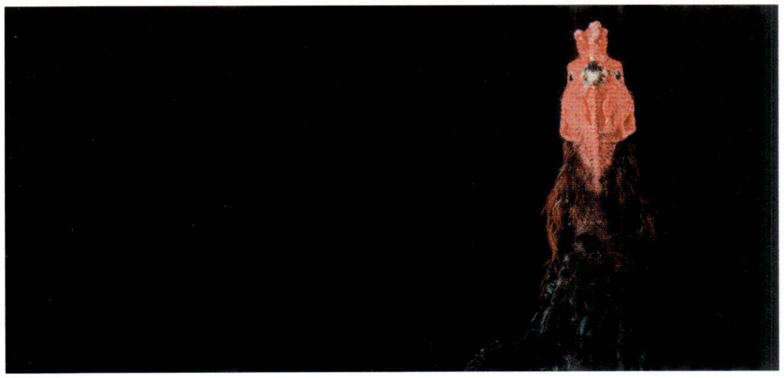

Thai game fowl (cock)

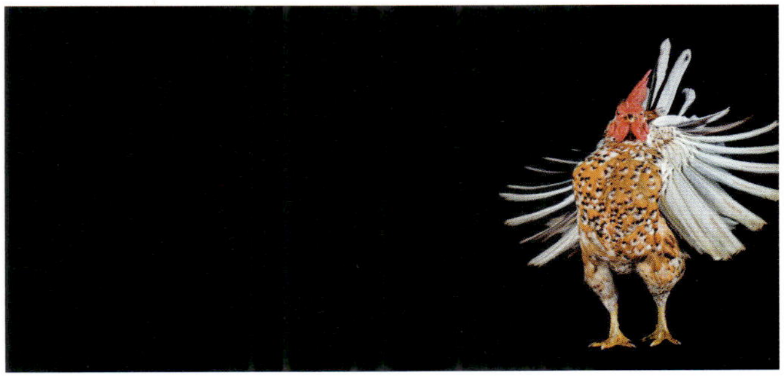

Serama (cock)

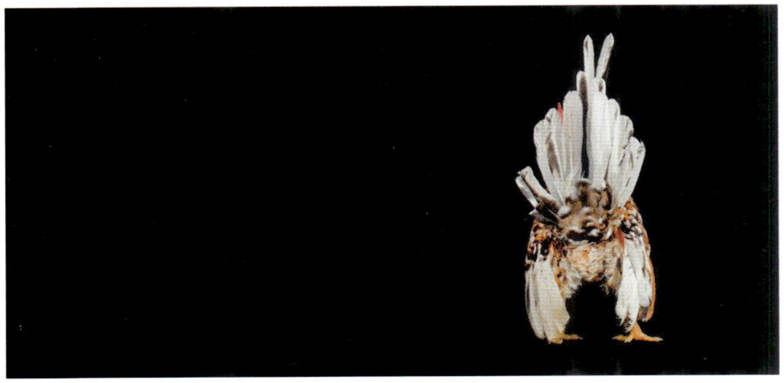

Serama (cock)

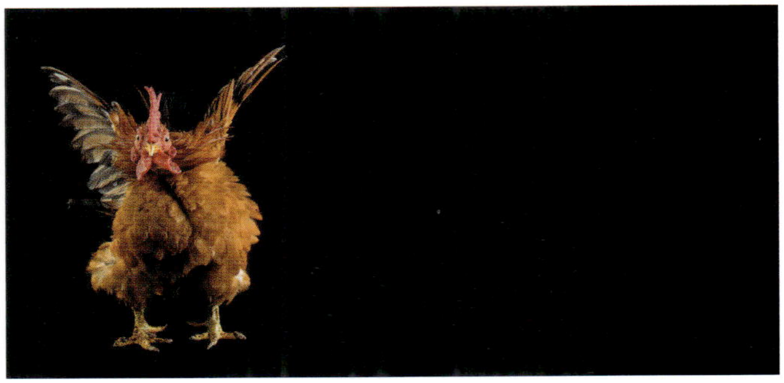

Serama (cock)

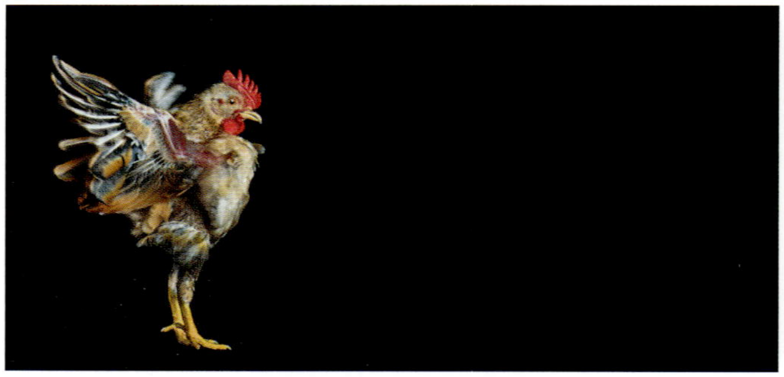

Serama (cock)

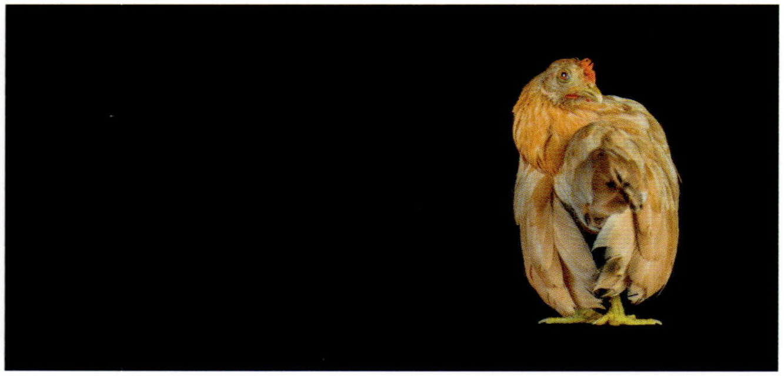

Serama (cock)

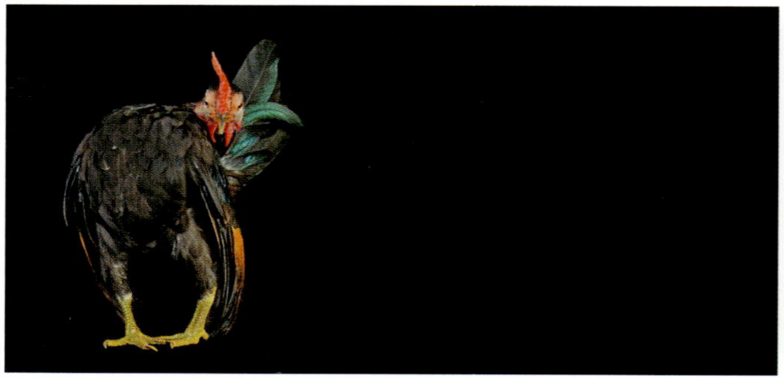

Serama (cock)

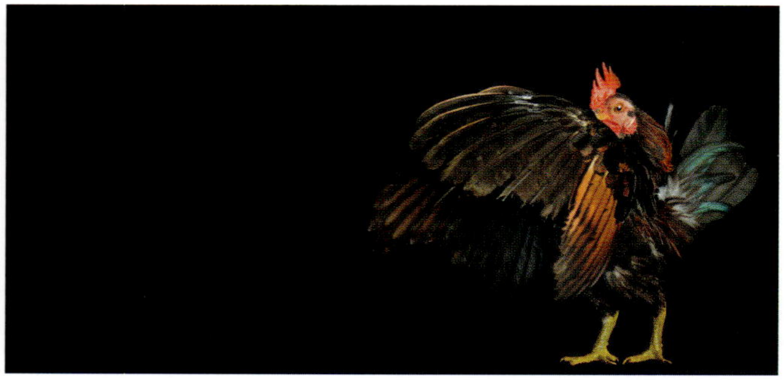

Serama (cock)

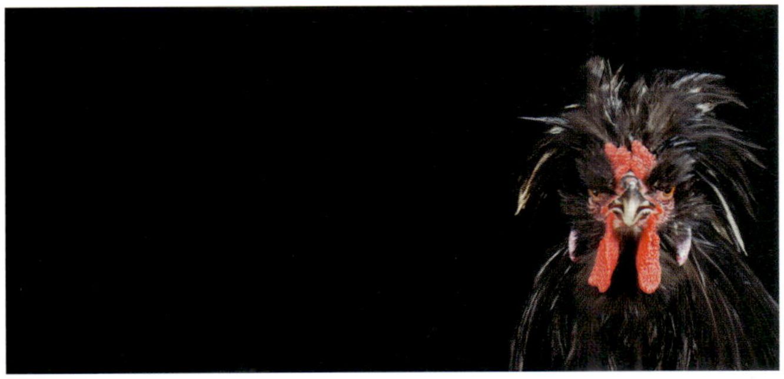

Black Sultan (cock)

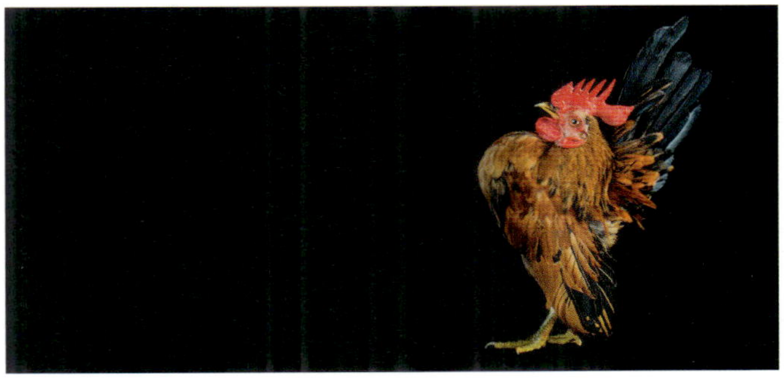

Serama (cock)

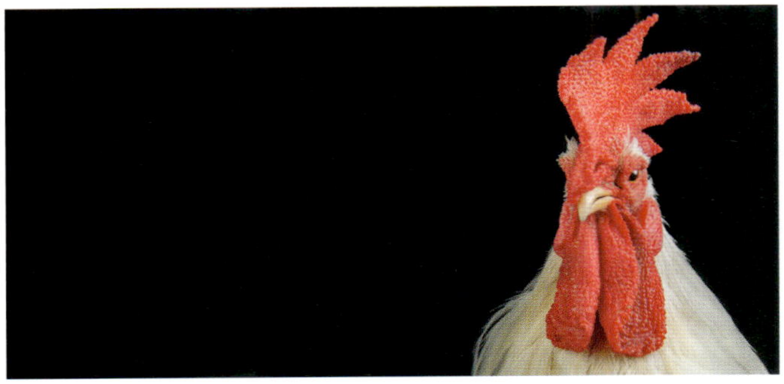

Japanese Bantam (cock)

Japanese Bantam (cock)

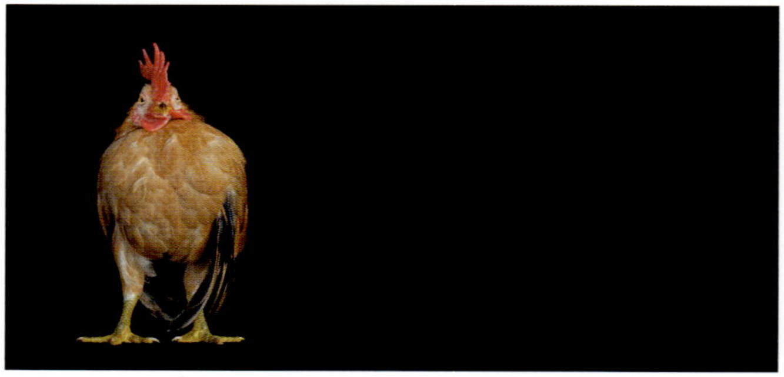

Serama (cock)

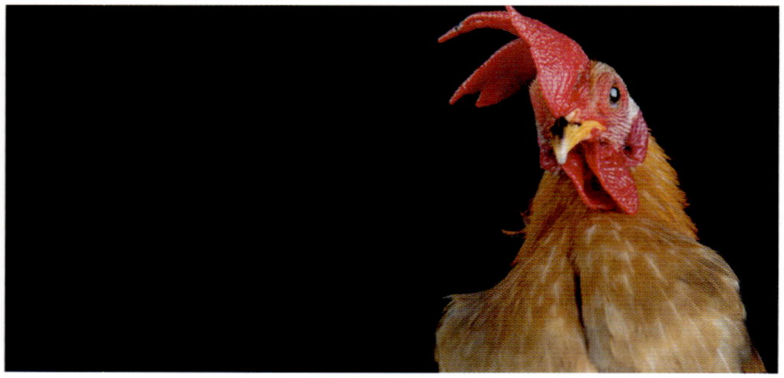

Serama (hen)

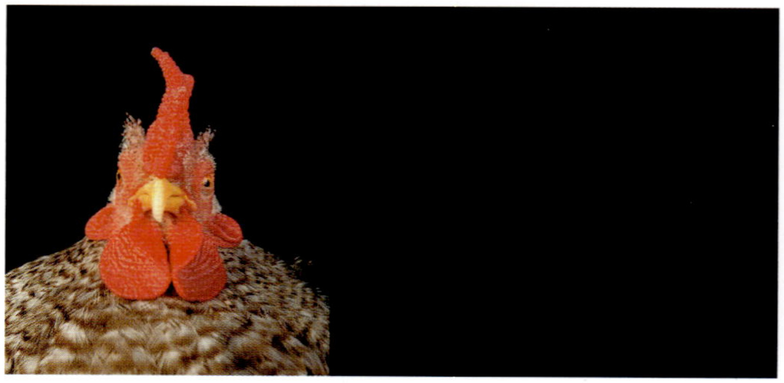

Serama (cock)

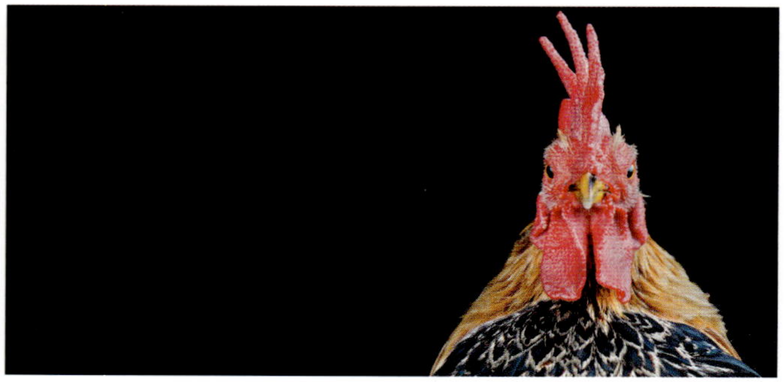

Serama (cock)

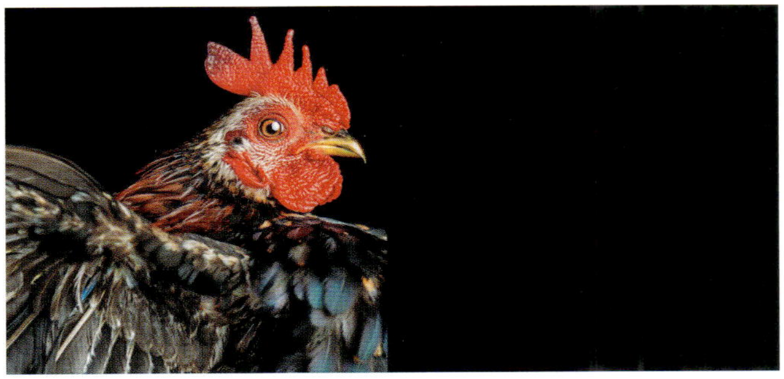

Serama (cock)

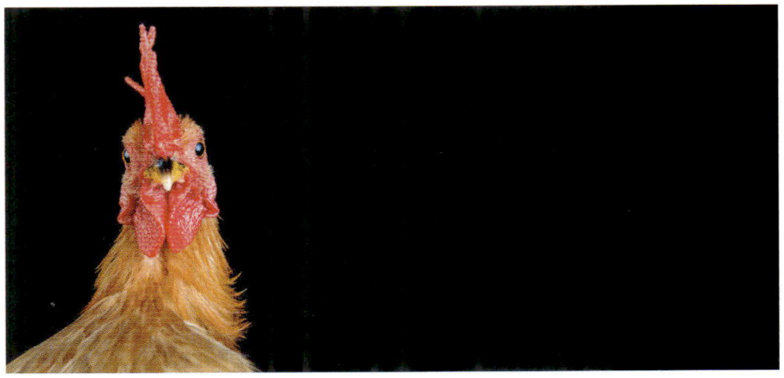

Serama (hen)

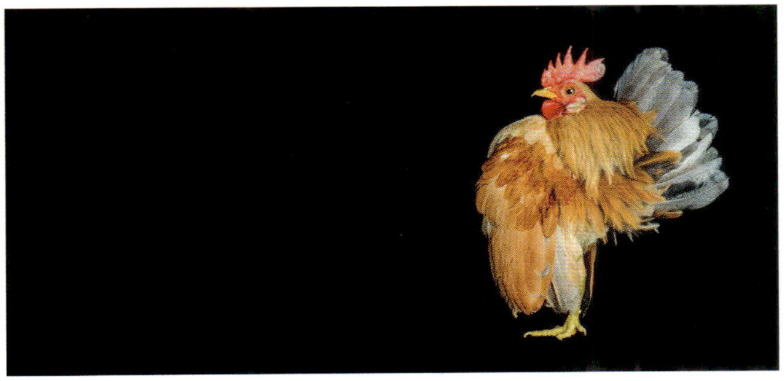

Serama (cock)

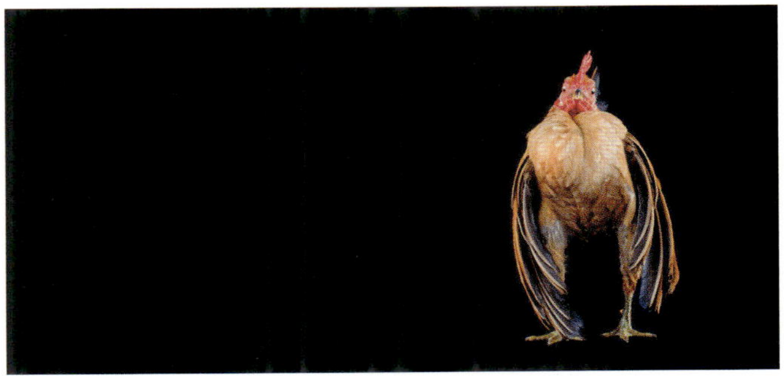

Serama (cock)

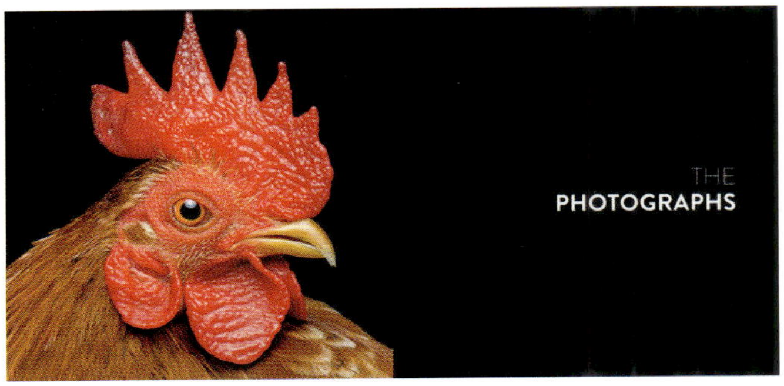

Serama (cock)

Serama (cock)

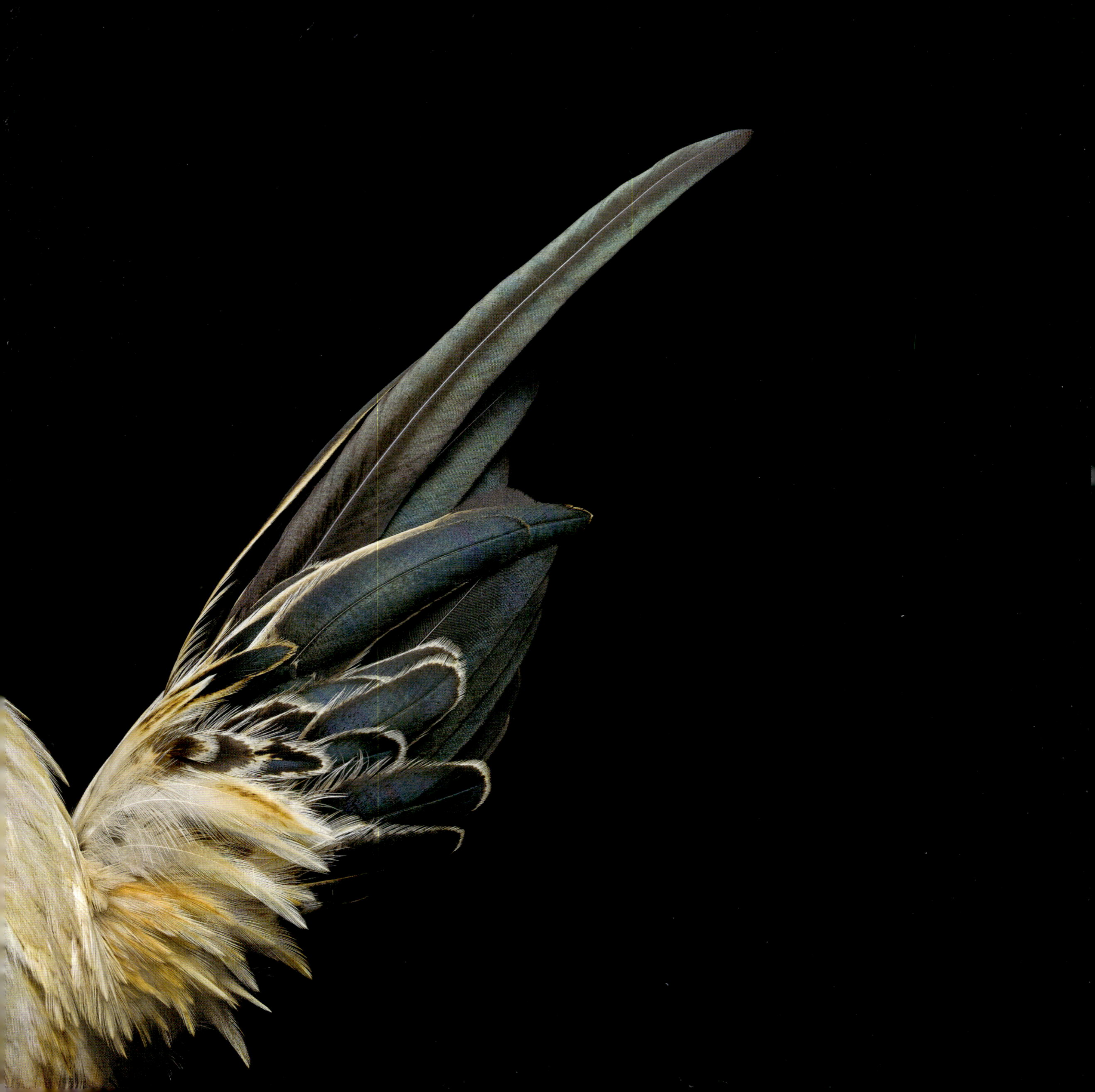

ACKNOWLEDGMENTS

Mohamad Tahar Jumaat, thank you for being the project's resident chicken handler, an able photographic assistant, and most of all a good friend as we zigzagged across Peninsular Malaysia. And to Rosmawati Moenir for being so supportive of our wild fowl chases.

Haqimin Salleh, you and your family have my gratitude for accommodating our request to photograph in your home at the very beginning of this project. Photographers are only able to do what we do because people like you open your doors to us.

Noppakun Suwanpayak and Nakamura Wong, thank you sharing your time, your wonderful community, and of course your enthusiasm for the Ayam Serama!

Pukku (Tuan Hassan), I truly appreciate your patience in taking the time to explain the finer points of the Ayam Serama culture. Your insights deepen my awareness not only of chickens, but of animals of all kinds.

My heartfelt gratitude goes to the following people who guided, inspired, and spurred me on during the journey of this work:

Agnès de Gouvion Saint-Cyr, Shirlene Noordin, Raphaël Millet and the staff of Phish Communications, Nina Gomiashvili, Edmund Wee and the staff of Epigram Books, Geoff Blackwell, Ruth Hobday, Bridget White, and Abby Aitcheson of PQ Blackwell, Chan Tse Chueen, Wesley Loh, Iskandar Jalil, Chow Chee Yong, Emmeline Yong, Denis Lim, Encik Matzi, Ian Singleton, Melvin Lee, Winfred Kwan, Chris Ang, Troy Toon, Wong Meng Fei, Richard Merrells, Caleb Ming, Bryan van der Beek, Tay Kay Chin, Darren Soh, Tow Heng Tan, Cathay Tan, Wendy Cheong, Charmaine Leung, Christopher Bailey, Judith Siu, Ore Huiying, Siong Chung Hua, and Jaclyn Teo.

For Abrams:
Editor: Anna Oseran

For PQ Blackwell:
Publisher: Geoff Blackwell
Editor in Chief: Ruth Hobday
Additional editorial: Abby Aitcheson
Book designer: Bridget White

Library of Congress Control Number: 2014940702

ISBN: 978-1-4197-1422-1

First edition published in 2015 by Abrams

Printed and bound in China

10 9 8 7 6 5 4 3 2 1

ABRAMS
THE ART OF BOOKS SINCE 1949
115 West 18th Street
New York, NY 10011
www.abramsbooks.com

The publisher is grateful for literary permissions to reproduce items subject to copyright. Every effort has been made to trace the copyright holders and the publisher apologizes for any unintentional omission. We would be pleased to hear from any not acknowledged here and undertake to make all reasonable efforts to include the appropriate acknowledgment in any subsequent editions.

He was like the cock who thought the sun had risen to hear him crow.
George Eliot

People who count their chickens before they are hatched act very wisely because chickens run about so absurdly it is impossible to count them accurately.
Oscar Wilde

If you can look a chicken in the eyes for ten seconds, it's yours forever.
From *Even Cowgirls Get the Blues* by Tom Robbins, published by Houghton Mifflin, 1976.

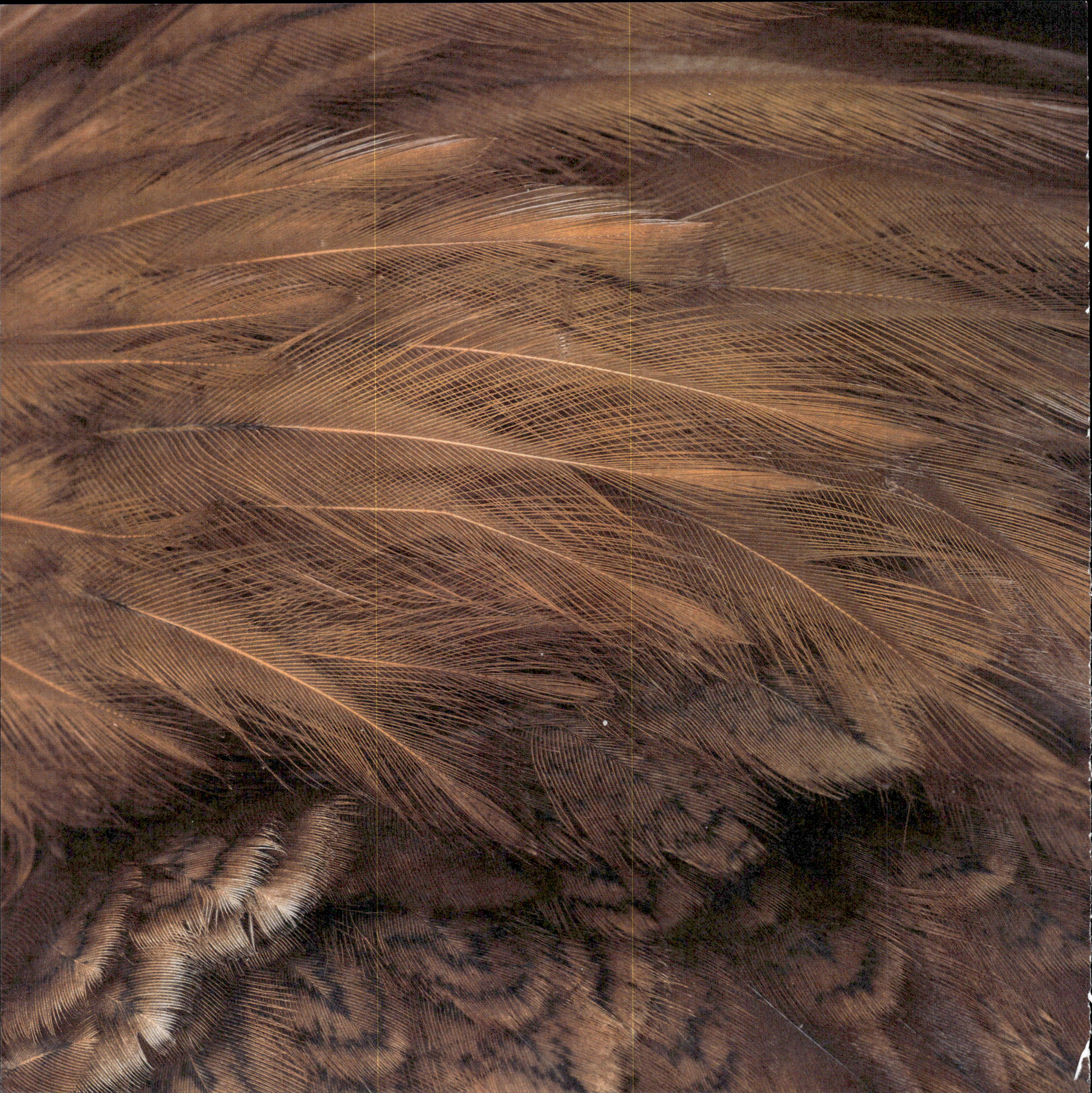